中国海洋统计年鉴

CHINA MARINE STATISTICAL YEARBOOK 2011

国 家 海 洋 局

Edited by

State Oceanic Administration,

People's Republic of China

海洋出版社

China Ocean Press

图书在版编目（CIP）数据

中国海洋统计年鉴. 2011 : 汉英对照 / 国家海洋局编著. -- 北京 : 海洋出版社, 2012.4
ISBN 978-7-5027-8250-4

Ⅰ. ①中… Ⅱ. ①国… Ⅲ. ①海洋－统计资料－中国－2011－年鉴－汉、英 Ⅳ. ①P7-66

中国版本图书馆CIP数据核字(2012)第074914号

责任编辑：王　溪

责任印刷：刘志恒

海洋出版社 出版发行

http: //www.oceanpress.com.cn

北京市海淀区大慧寺路 8 号　　邮编：100081

国家海洋信息中心印刷厂印刷　　新华书店发行所经销

2012 年 5 月第 1 版　　2012 年 5 月第 1 次印刷

开本：787mm×1092mm　1/16　印张：19.25

字数：495 千字

定价：168.00 元

发行部：(010)62147016　　邮购部：(010)68038093　　总编室：(010)62114335

《中国海洋统计年鉴》编委会

柯王俊	中国船舶工业集团公司
赵宗波	中国船舶重工集团公司
李　军	中国石油天然气集团公司
刘　岩	中国石油化工集团公司
陈伟杰	中国海洋石油总公司
胡红江	中国盐业总公司
王华俊	中国有色金属工业协会
杨宝瑞	辽宁省海洋与渔业厅
王保民	河北省国土资源厅
孙连友	天津市海洋局
王守信	山东省海洋与渔业厅
唐庆宁	江苏省海洋与渔业局
沈依云	上海市海洋局
陈宗尧	浙江省海洋与渔业局
张福寿	福建省海洋与渔业厅
洪伟东	广东省海洋与渔业局
杨小光	广西壮族自治区海洋局
潘建纲	海南省海洋与渔业厅
栾玉瑄	大连市海洋与渔业局
黄聿颂	青岛市海洋与渔业局
吴建义	宁波市海洋与渔业局
王春生	厦门市海洋与渔业局
梁俊乾	深圳市海洋局

《中国海洋统计年鉴》编辑部

主　任：

何广顺　　国家海洋信息中心副主任

副主任：

沈　君　　国家海洋局政策法规和规划司海洋经济处处长

王晓惠　　国家海洋信息中心海洋经济部主任

成　员：

王占坤　杨　娜　李长如　郭　越　宋维玲　赵　锐

林香红　董　伟　周洪军　蔡大浩　赵　鹏　朱　凌

徐丛春　李宜良　周怡圃　刘　彬　张潇娴（国家海洋信息中心）

冀渺一　庞　冲（国家海洋局政策法规和规划司海洋经济处）

特邀编辑：

李艳丽　　教育部

秦浩源　　科技部

苗　强　　国土资源部

蒋　琢　　交通运输部

魏新平　　水利部

郭　睿　　农业部

毛玉如　　环境保护部

刘建杰　　国家林业局

周　鲲　　国家旅游局

崔胜先　　中国科学院

陆覃星　　中国地震局

吴万克　　中国气象局

裘音浪　　中国船舶工业集团公司

Cao Ning	Chinese Academy of Sciences
Xu Tieju	China Earthquake Administration
Wang Bangzhong	China Meteorological Administration
Ke Wangjun	China State Shipbuilding Corporation
Zhao Zongbo	China Shipbuilding Industry Corporation
Li Jun	China National Petroleum Corporation
Liu Yan	China Petrochemical Group Corporation
Chen Weijie	China National Offshore Oil Corporation
Hu Hongjiang	China National Salt Industry Corporation
Wang Huajun	China Nonferrous Metals Industry Association
Yang Baorui	Department of Ocean and Fisheries of Liaoning Province
Wang Baomin	Hebei Province Department of Land and Resources
Sun Lianyou	Tianjin Ocean Administration
Wang Shouxin	Department of Oceanic and Fishery of Shandong Province
Tang Qingning	Jiangsu Provincial Ocean and Fisheries Bureau
Shen Yiyun	Shanghai Ocean Administration
Chen Zongyao	Zhejiang Provincial Ocean and Fisheries Bureau
Zhang Fushou	Department of Oceanic and Fishery of Fujian Province
Hong Weidong	Guangdong Provincial Oceanic and Fishery Administration
Yang Xiaoguang	Guangxi Ocean Administration
Pan Jiangang	Department of Marine and Fishery of Hainan Province
Luan Yuxuan	Dalian Ocean and Fishery Administration
Huang Yusong	Qingdao Ocean and Fishery Administration
Wu Jianyi	Ningbo Ocean and Fishery Bureau
Wang Chunsheng	Oceans and Fisheries Bureau of Xiamen
Liang Junqian	Shenzhen Ocean Administration

Editorial Department of
China Marine Statistical Yearbook

编 者 说 明

一、《中国海洋统计年鉴》2011 年版是一部全面反映 2010 年中华人民共和国海洋经济发展和海洋管理服务情况的资料性年鉴，全书为中英文对照。

二、本年鉴的统计资料范围为人们在海洋和沿海地区开发、管理、利用海洋资源和空间，发展海洋经济的生产和活动以及沿海地区的社会经济概况。地域范围为沿海地区、沿海城市和沿海地带，其排列顺序按《沿海行政区域分类与代码》(HY/T 094-2006) 的顺序排列。

三、本年鉴内容包括综合资料、海洋经济核算、主要海洋产业活动、主要海洋产业生产能力、涉海就业、海洋科学技术、海洋教育、海洋环境保护、海洋行政管理及公益服务、全国及沿海社会经济、部分世界海洋经济统计资料等十一部分。

四、本年鉴根据《海洋统计报表制度》（国统制[2009]42 号）和《海洋生产总值核算制度》（国统制[2011]7 号），资料主要来源于沿海省、自治区、直辖市统计局、海洋厅（局）以及 20 个有关涉海部、局、总公司。

五、本年鉴中除特殊说明外，所有价值量指标均为当年价，除注明年份外，其他均为 2010 年数据，年鉴中每部分后附有主要海洋统计指标解释，对主要海洋统计指标的含义、统计范围和统计方法做了简要说明。统计数据中的其他说明置于表的下面。有续表的资料，如有注释均置于最后一张表的下面。

六、本年鉴表格中符号使用说明:“…”表示数据不足本表最小计算单位数；“空格”表示该项统计指标数据不详或无该项数据；“#”表示其中的主要项；其他符号如“*”或“①”等表示本表后面有注释。

七、本年鉴中由于数字精确度的原因，四舍五入后部分分项值之和与合计值有微小差异。

八、本年鉴资料国内部分未包括香港特别行政区、澳门特别行政区和台湾省数据。

九、《中国海洋统计年鉴》在编汇过程中，得到了各有关单位的大力支持，我们在此表示衷心的感谢。本年鉴中如有疏漏和不妥之处，敬请读者批评指正。

《中国海洋统计年鉴》编辑部

Introduction

I. *China Marine Statistical Yearbook (2011)* is a data almanac which reflects in an all-round way the development of marine economy, marine management and service in the People's Republic of China in 2010, and it is a Chinese-English bilingual edition.

II. The Yearbook's statistics cover the production and activities in the marine and coastal areas in relation to the development, management and utilization of marine resources and space, and the development of marine socioeconomy. The regions covered are the coastal regions, coastal cities and coastal zones with coastlines, which are arranged in order according to the *Coastal Administrative Areas Classification and Codes* (HY/T 094—2006).

III. The data in the yearbook consist of 11 sections, namely, integrated data, marine economic accounting, major marine industrial activities, production capacity of major marine industries, ocean-related employment, marine science and technology, marine education, marine environmental protection, marine administration and public-good service, national and coastal socioeconomy, part of the world's marine economic statistics data.

IV. The Yearbook is based on the *Marine Statistics Report System* (Guotongzhi [2009] No. 42) and the *Ocean Gross Product Accounting System* (Guotongzhi [2011] No. 7), its data mainly come from the statistical bureaus and the oceanic administrations of the coastal provinces, autonomous regions, and municipalities directly under the Central Government as well as the 20 ocean-related ministries, bureaus and general corporations concerned.

V. Unless otherwise specified in the Yearbook, all the value indicators are given at the current price. Each section is attached by explanatory notes to the major marine statistical indicators, giving a brief explanation for the meaning, statistical range and statistical methods of the major marine statistical indicators. Other notes to the statistical data are listed below the tables. For the data with continued tables, annotations, if any, are put below the last table.

VI. The usage of symbols in the tables: "…" indicates the statistics smaller than the minimum calculation unit in the table; "Blank" indicates that the data of the statistical index is unknown for the time being or that there shouldn't be any data; "#" indicates the major items of the table; Other symbols, such as "*" or "①", indicate "see footnotes below".

VII. For reasons of digital accuracy, there is small difference between the sum of values of some subterms after having been rounded off and the total values.

VIII. The domestic part of the Yearbook does not include that of Hong Kong Special Administrative Region, Macau Special Administrative Region and Taiwan Province.

IX. In the course of editing *China Marine Statistical Yearbook*, we enjoyed energetic support from the various departments concerned and we hereby extend our heartfelt thanks to them. Criticisms and comments are welcome from readers on any of the oversights and inappropriateness as the time for editing is too short.

Editorial Department of
China Marine Statistical Yearbook

目　次
CONTENTS

7 海洋教育

Marine Education

8 海洋环境保护

Marine Environmental Protection

9 海洋行政管理及公益服务
Marine Administration and Public-Good Service

2010年我国海洋经济发展综述

2010年，各级海洋行政主管部门深入贯彻实践科学发展观，积极落实国家“保增长、调结构”的政策方针，加快海洋经济结构调整，促进海洋经济发展方式转变，努力克服国际金融危机等不利影响，全国海洋经济实现平稳发展。

一、全国海洋经济发展概况

2010年，全国海洋生产总值39 572.7亿元，比上年增长14.7%（除特别注明外，增长率均按可比价计算），海洋生产总值占国内生产总值的9.86%，占沿海地区生产总值的16.1%。全国涉海就业人员3 350.8万人，比上年增长80万人。

二、主要海洋产业发展情况

2010年，主要海洋产业实现增加值16 187.8亿元，比上年增长17.4%，占海洋生产总值的40.9%，滨海旅游业和海洋交通运输业仍占主导地位。

海洋第一产业 2010年，沿海地区海洋渔业生产平缓增长，全年海洋渔业实现增加值2 852亿元，比上年增长6.0%。海水产品产量达2 797.5万吨，比上年降低2.9%，其中，海水养殖产量1 482.3万吨，比上年降低3.5%；海洋捕捞产量1 203.6万吨，比上年降低10.5%；远洋捕捞产量88.8万吨，比上年增长8.5%。

海洋第二产业 随着海洋风电进入规模开发阶段，海洋电力业继续保持高速增长，全年实现增加值38.1亿元，比上年增长80.1%。海洋工程建筑业实现增加值874.2亿元，比上年增长23.9%。在国家一系列政策和《船舶工业调整与振兴规划》的指导下，海洋船舶工业攻坚克难，继续保持平稳快速发展，造船完工量及新承接船舶订单量大幅增长，全年实现增加值1 215.6亿元，比上年增长22.8%。海洋化工业全年实现增加值613.8亿元，比上年增长22.1%。随着管理力度的加强，我国海砂开采活动更加规范有序，海洋矿业，全年实现增加值45.2亿元，比上年减少8.8%。我国海水利用业相关政策措施逐步完善，产业化水平进一步提升，继续保持较快发展，全年实现增加值8.9亿元，比上年增长9.0%。海盐产量虽受天气影响略有下降，但市场行情看涨，海洋盐业仍实现了良好的经济增长，全年实现增加值65.5亿元，比上年增长13.7%，产盐量为3 286.63万吨，比上年降低6.1%。海洋生物医药业全年实现增加值83.8亿元，比上年增长29.9%。海洋油气业全年实现增加值1 302.2亿元，比上年增长53.9%，全年海洋原油产量4 710.0万吨，比上年增长27.4%，海洋天然气产量1 108 905万立方米，比上年增长29.1%。

海洋第三产业 2010年，随着国际贸易形势趋好和航运价格恢复性增长，海洋交通运输业迅速回暖，全年实现增加值3 785.8亿元，比上年增长20.6%，沿海港口货物吞吐量达564 464万吨，比上年增长15.8%，国际标准集装箱吞吐量13 145万标准箱，比上年增长19.3%。滨海旅游业全年实现增加值5 303.1亿元，比上年增长13.3%。

三、区域海洋经济发展情况

2010年，沿海各地认真落实党中央、国务院发展海洋经济的战略部署，不断推进海洋经济结构调整，有效巩固和扩大应对国际金融危机成果，区域海洋经济总量迅速增长，环渤海经济区、长江三角洲经济区和珠江三角洲经济区海洋生产总值分别为13 868.5亿元、12 658.9亿元和8 253.7亿元，分别占全国海洋生产总值35.0%、32.0%和20.9%。

环渤海经济区海洋经济增长迅速，海洋生产总值比上年增长24%(现价)，占地区生产总值比重达15.9%。海洋产业增加值为7 925.3亿元，海洋相关产业增加值为5 943.3亿元。与上年相比各项主要海洋产业均实现增长，海洋渔业、海洋交通运输业、滨海旅游业三大海洋支柱产业增加值合计达到4 483.3亿元，占该地区主要海洋产业增加值的71.2%。海洋生物医药业、海洋电力业、海水利用业等海洋新兴产业增幅明显，与上年相比分别增长47.5%、96.7%、63.0%(现价)。受国际原油市场成品油价大幅增长以及渤海新投产油气田影响，海洋油气业增幅远超去年，其增加值翻了一倍多，增速达到128.1%(现价)。

长江三角洲经济区海洋经济保持较快增长，2010 年海洋生产总值比去年增长 22.7%（现价），海洋生产总值占地区生产总值比重 14.7%。长江三角洲海洋产业增加值 7 172.1 亿元，海洋相关产业增加值 5 486.8 亿元。按照产值贡献，滨海旅游业、海洋交通运输业、海洋船舶工业和海洋渔业四个产业位居前列，其增加值之和占该地区主要海洋产业增加值的 92.5%，其中滨海旅游业得益于上海世博会的召开，产值贡献位居第一。按照增长速度，海洋油气业增长最为迅猛，增长速度达 285.7%（现价），海洋化工业、海洋电力业、海洋生物医药业等海洋新兴产业也呈现出快速增长态势，增长速度分别达到 77.9%、73.9%和 76.9%（现价）。

2010 年，珠江三角洲经济区海洋经济快速增长，继续发挥着海洋经济发展的主力军作用。2010 年，珠江三角洲海洋生产总值达 8 253.7 亿元，比上年增长 23.9%（现价），海洋生产总值占地区生产总值比重达 17.9%。海洋产业增加值为 5 066.0 亿元，海洋相关产业增加值为 3 187.7 亿元。滨海旅游业、海洋交通运输业、海洋油气业、海洋化工业和海洋渔业依然为珠江三角洲经济区海洋经济发展的支柱产业，其增加值之和占该地区主要海洋产业增加值 92.2%。海洋油气业、海洋电力业和海洋工程建筑业增长迅速，增长速度分别达到 85.6%、55.7% 和 57.3%（现价）。

四、海洋科研教育管理服务情况

2010 年，海洋科研教育事业继续快速发展，海洋科研机构共 181 个，从业人员 35 405 人，海洋科研项目 13 466 项，发表海洋科技论文 14 296 篇；出版海洋科技著作 254 种。开设海洋专业的高等学校共 327 个，专任教师 239 655 人，高等教育和中等职业教育海洋专业毕业生 91 852 人，招生 136 663 人，在校生 371 260 人，毕业班学生 111 202 人，海洋教育为海洋事业的蓬勃发展培养了大批人才。

2010 年，我国海洋环境质量总体维持在较好水平，主要海洋功能区环境质量基本满足海域使用要求，海洋赤潮和绿潮灾害有所减轻；我国管辖海域的海水环境质量状况总体较好，11 个沿海地区工业废水排放总量为 1 413 763.8 万吨，比上年增长 1.8%，工业废水排放达标率为 96.9%，比上年提高 0.8 个百分点；海洋保护区环境状况总体良好，主要保护对象或保护目标基本保持稳定。我国近岸典型海洋生态系统总体处于健康和亚健康状况，90%监控区域的生态系统基本维持其自然属性，全国沿海对 18 个海洋生态监控区开展了监测，处于健康、亚健康和不健康状态的海洋生态监控区分别占 14%、76%和 10%；对 100 个入海排污口邻近海域水质、沉积物质量和生物质量的监测结果显示，入海排污口邻近海域环境质量总体状况仍然较差，与上年相比未见明显改善。我国管辖海域的海水环境质量状况总体较好，符合第一类海水水质标准的海域面积约占我国管辖海域面积的 94%，但尚有 4.8 万平方千米的近岸海域水质劣于第四类海水水质标准。全年由于风暴潮灾害导致受灾人口 437.07 万人，比上年减少近一半，直接经济损失为 65.79 亿元，比上年减少了 22.6%；全海域共发现赤潮 69 次，累计面积约 10 892 平方千米，与上年基本持平。

2010 年，海洋行政管理工作开展有序。全年颁发海域使用权证书 2 481 本，确权海域面积 19.4 万公顷；全年共签发疏浚物海洋倾倒许可证 322 份，实施各项海洋执法检查共 72 233 次，发现违法行为 1 836 起，比上年略有增加；提供海洋数值预报服务共 17 683 次，开展海洋调查项目 4 230 项，获得数据共 41 640 371 个，各项海洋监测获得数据量共 24 211.2 万个，全年接收存档卫星遥感数据量共计 28 773.0 GB；室（馆）存海洋档案案卷 118 334 卷（册），磁介质档案 8 992 盘，全年共接待读者 2 814 人次。

Summary of National Marine Economic Development in 2010

In 2010, the competent marine administrative departments at all levels make an in-depth study of and put into practice the concept of scientific development, actively carry out the policy of "ensuring growth and adjusting structures", speed up the marine economic structure adjustment and the development mode change, make great effort to conquer such unfavorable impacts as the international financial crisis, and give an impetus to the stable development of national marine economy.

I. Survey of the National Marine Economic Development

In 2010, the gross ocean product of the national marine economy reaches 39 57.27 billion Yuan, up 14.7% from the previous year (Unless otherwise specified, the growth rate is calculated in the comparable price.), accounting for 9.86% of the GDP, and 16.1% of the coastal Gross Regional Product. The number of people employed by the ocean-related sectors reaches 33.508 million, increasing by 0.8 million from the previous year.

II. Development Situation of Major Marine Industries

In 2010, major marine industries effect an added value of 1 618.78 billion Yuan, up 17.4% from the previous year, accounting for 40.9% of the GOP, and the coastal tourism and marine communications and transportation industry still play a dominant role.

Primary marine industry In 2010, the marine fishery production of the coastal region increases gently. The full-year marine fishery effects an added value of 285.2 billion Yuan, up 6.0% from the previous year; the output of seawater products amounts to 27.975 million tons, registering a decrease by of 2.9% as compared with that of the previous year, among which, the production of mariculture amounts to 14.823 million tons, 3.5% down from the previous year; the yield from marine fishing 12.036 million tons, 10.5% down from the previous year; and that from the deep-sea fishing 0.888 million tons, 8.5% up from the previous year.

Secondary marine industry With the coastal wind power steping into the stage of large-scale development, marine electric power industry continues to keep a high speed increase, and effects an added value of 3.81 billion Yuan for the whole year，increasing by 80.1% from the previous year. The marine engineering architecture effects a full-year added value of 87.42 billion Yuan, 23.9% up from the previous year. The marine shipbuilding industry has over come many difficulties and continues to keep a stable and rapid development under the guidance of the Shipbuilding Industry Restructuring and Revitalization Plan, the completed quantity of ships built and the number of nearly received shipbuilding orders have increased by a large margin, and the full-year added value reaches 121.56 billion Yuan, increasing by 22.8%. The marine chemical industry accomplishes an added value of 61.38 billion Yuan for the whole year, 22.1% up from the previous year. With the improved management, China's sea sand exploitation activities become more normal and orderly, and the marine mining industry effects a full-year added value of 4.52 billion Yuan, 8.8% down from the previous year. The improvement of the related policies and measures and the enhancement of the industrialization level make the seawater utilization industry keep a high speed increase, and the sea water desalination and multipurpose utilization industry effects a full-year added value of 890 million Yuan, 9.0% up from the previous year. Sea salt production has declined slightly affected by the weather, but the market is bullish, the marine salt industry still achieved good economic growth, with an annual added value of 6.55 billion yuan, an increase of 13.7 percent over the previous year, and the production of salt 32.9 million tons, 6.1% lower than last year. The marine biomedicine industry accomplishes an added value of 8.38 billion Yuan for the whole year, 29.9% up from the previous year. The offshore oil and gas industry effects a full-year added value of 130.22 billion Yuan, 53.9% up from the previous year. The annual crude oil output is 47.1 million tons, 27.4% up from the previous year, and the output of marine natural gas amounts to 11.09 billion cubic metres, 29.1% up from the previous year.

Tertiary marine industry In 2010, as the international trade situation is getting better and the shipping prices has resumed their growth, the marine communications and transportation industry has got warm quickly, and an added value of 378.58 billion Yuan is accomplished for the whole year, 20.6% up from the previous year. The cargo handling capacity of coastal harbors amounts to 5 644.64 million tons, 15.8% up from the previous year and the handling capacity of international standardized containers 131.45 million standard cases, 19.3% up from the previous year. The coastal tourism effects a full-year added value of 530.31 billion Yuan, 13.3% up from the previous year.

III. Situation of Regional Marine Economic Development

In 2010, as coastal regions earnestly implement the marine economy development strategy of the Party Central Committee and the State Council, and adjust the marine economy structure, consolidate and expand the results of dealing with the international financial crisis, China's regional marine economy has grown rapidly, the gross ocean products of the Round-the-Bohai Economic Zone, Changjiang River Delta Economic Zone and Zhujiang River Delta Economic Zone are 1 386.85 billion Yuan, 1 265.89 billion Yuan and 825.37 billion Yuan respectively, accounting for 35.0%, 32.0% and 20.9% of the national gross ocean product, respectively.

The marine economy in the Round-the-Bohai Sea Economic Zone grows rapidly, the Gross Ocean Product increases by 24% as against that last year(at the current price), and accounts for 15.9% of the Gross Regional Product. The added value of marine industries is 792.53billion Yuan and that of ocean-related industries 594.33 billion Yuan. The major marine industries have grown, The added value of the three major marine pillar industries, i.e., marine fishery ,marine communications and transportation industry, and coastal tourism, amounts to 448.33 billion Yuan, accounting for 71.2%of the added value of major marine industries in the region. The growth rates of the marine new industries such as marine biomedicine, marine electric power and seawater initialization are obvious, 47.5%，96.7% and 63.0%(at the current price), respectively. Affected by the sharp rise of the international oil prices and the new oil and gas fields put into production in the Bohai Sea, the increase in the offshore oil and gas industry is far more than that last year, the added value of it is more than doubled, and the growth rate is 128.1%(at the current price).

The growth rate of the marine economy in the Changjiang River Delta Economic Zone continues to grow rapidly. In 2010, the Gross Ocean Product increases by 22.7% (at the current price), and its proportion in the Gross Regional Product is 14.7%. The added value of marine industries is 717.21 billion Yuan and that of ocean-related industries 548.68 billion Yuan. In terms of their contributions to the output value the industries of coastal tourism, marine communications and transport, marine shipbuilding, and marine fishery rank among the first, and the added value of these industries accounts for 92.5% of the added value of major marine industries in the region. Among them，due to the convening of the Shanghai World Expo, the coastal tourism industry ranks first. According to the growth rate, the offshore oil and gas industry grows most rapidly, at a growth rate of 285.7% (at the current price), and new industries such as marine chemical, marine electric power and marine biomedicine industries present a posture of rapid growth ,the growth rates of these industries being 77.9%,73.9% and 76.9% (at the current price), respectively.

In 2010, the marine economy in the Zhujiang River Delta Economic Zone continues to grow rapidly and becomes the main force leading the rapid development of marine economy. In 2010, the gross ocean product of the Zhujiang River Delta Economic Zone reaches 825.37 billion Yuan, 23.9% (calculated at the current price) up from the previous year, occupying a proportion of 17.9% in the Gross Regional Product. The added value of marine industries is 506.60 billion Yuan and that of the ocean-related industries 318.77 billion Yuan. The coastal tourism, marine communications and transportation, marine chemical industry, offshore oil and gas industry, and marine fishery, are still the pillar industries of marine economic development in the zone, and the added value of these industries accounts for 92.2%(calculated at the current price) of the added value of major marine industries in the region. And the growth rates of offshore oil, marine electric power and marine engineering architecture are 85.6%, 55.7% and 57.3% (at the current price), respectively.

IV. Marine Scientific Research Education and Public Service

In 2010, marine scientific research and education keep on developing. The number of marine scientific research institutions is 181, and they employ 35 405 people. The number of marine scientific research projects reaches 13 466, and 14 296 marine scientific and technological papers and 254 kinds of marine scientific and technological works have been published. There are 327 ordinary marine colleges and universities that set up marine specialties, employing 239 655 full-time teachers. In the marine specialties of ordinary higher education and secondary vocational education there are 91 852 graduates, 371 260 at school, 136 663 new students and 111 202 in the graduating classes. Marine education has trained a large number of talents for the development of national marine undertakings.

In 2010, China's marine environmental quality generally maintains a fairly good level, the environmental quality of the major marine functional zones can basic meet the requirement for sea area use and the marine red tide and green tide disasters are somewhat mitigated; The condition of seawater environmental quality of the sea areas under China's jurisdiction is on the whole good, the total discharge of industrial waste water from the 11 coastal regions is 14.137 638 billion tons, 1.8% up from the previous year and the up-to-standard discharge rate of industrial waste water is 96.9%, 0.8 percentage points up from last year; The environmental condition of marine reserves is good on the whole and the main objects of protection or objectives for protection basically keep stable. China's nearshore typical marine ecosystems are in generally in a healthy or sub-healthy state and the ecosystems in the 90% of the monitored areas basically maintain their natural attributes. Eighteen marine ecological monitoring areas in the coastal repair of China have been monitored and the marine ecological monitoring areas that are in the healthy, sub-healthy or unhealthy state occupy 14%, 76% and 10%, respectively. The monitoring results for the water, sediment and biological quality of the sea areas near 100 drainage exits into the sea indicate that the overall condition of environmental quality there is still poorer, with no obvious improvement as compared with the previous year. The condition of seawater environmental quality of the sea area under China's jurisdiction is good on the whole and the area of the sea which meets the standard of the class I seawater quality accounts for about 94% of China's jurisdictional sea area, but there is still 48 000 km^2 nearshore sea area whoes water quality is poorer than the class IV seawater quality. The population afflicted by the storm surge disaster throughout the year reaches 4.370 7 million, nearly half less than the last year and the direct economic is 6.579 billion Yuan, 22.6% less than the previous year; A total of 69 red tides and found across the whole sea area, covering about 10 892 km^2 in the aggregate, basically balancing with the previous year.

In 2010, marine administration is carried out in an orderly way. A total of 2 481 certificates for the sea area use are issued for the whole year, the sea area with the ownership of patent rights having reached 193 769 hm^2; A total of 322 permits for oceanic dumping of dredged material are signed and issued, and various marine inspections for law enforcement are carried out on 72 233 occasions, in which 1 836 cases of unlawful practice are discovered, a little higher than the previous year; Marine numerical forecast service is provided on 17 683 occasions, 4 230 marine survey items are carried out and 41 640 371 data are acquired, the data quantity obtained from various items of marine monitoring is 242.112 million and the satellite remote-sensing data received and placed on file total 28 773.0 GB; The number of marine files kept in the divisions (Marine Archives) reaches 118 334 volumes (copies), with 8 992 disks of magnetic-medium file. For the whole year, the number of readers received amounts to 2 814 person-times.

图1 全国海洋生产总值及三次产业构成

China's Gross Ocean Product and Three Industries Composition

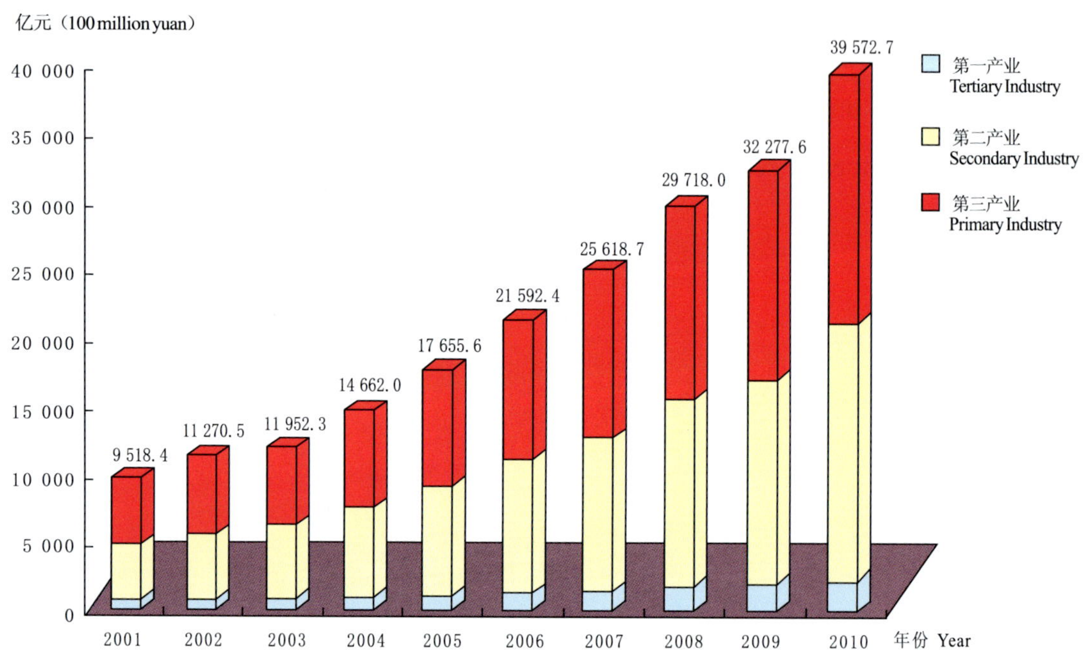

图2 2010年全国主要海洋产业增加值构成

Composition of Added Values of the Major Marine Industries in China in 2010

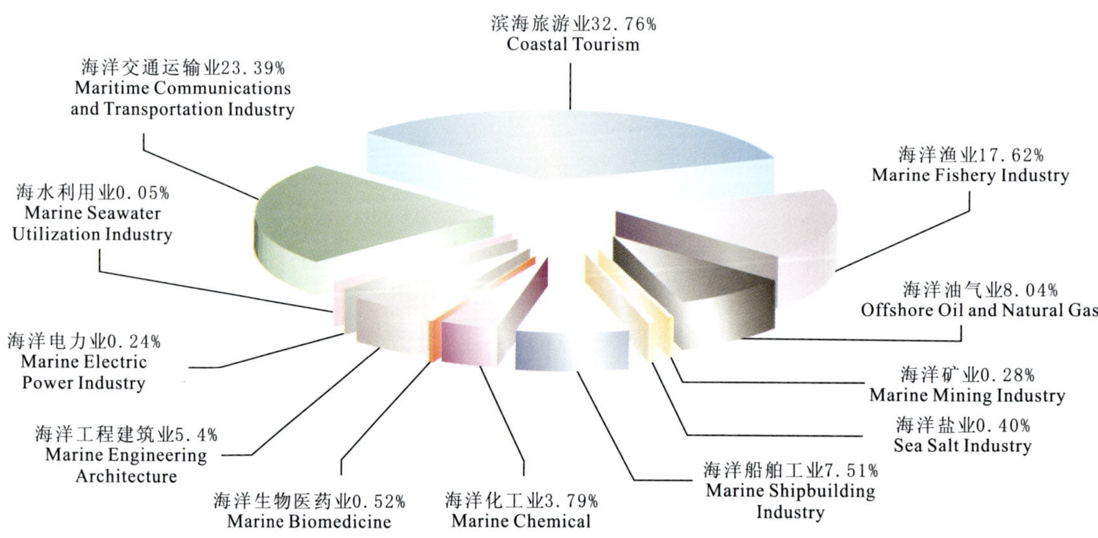

图3　2010年沿海地区海洋生产总值
Gross Ocean Product by Coastal Regions in 2010

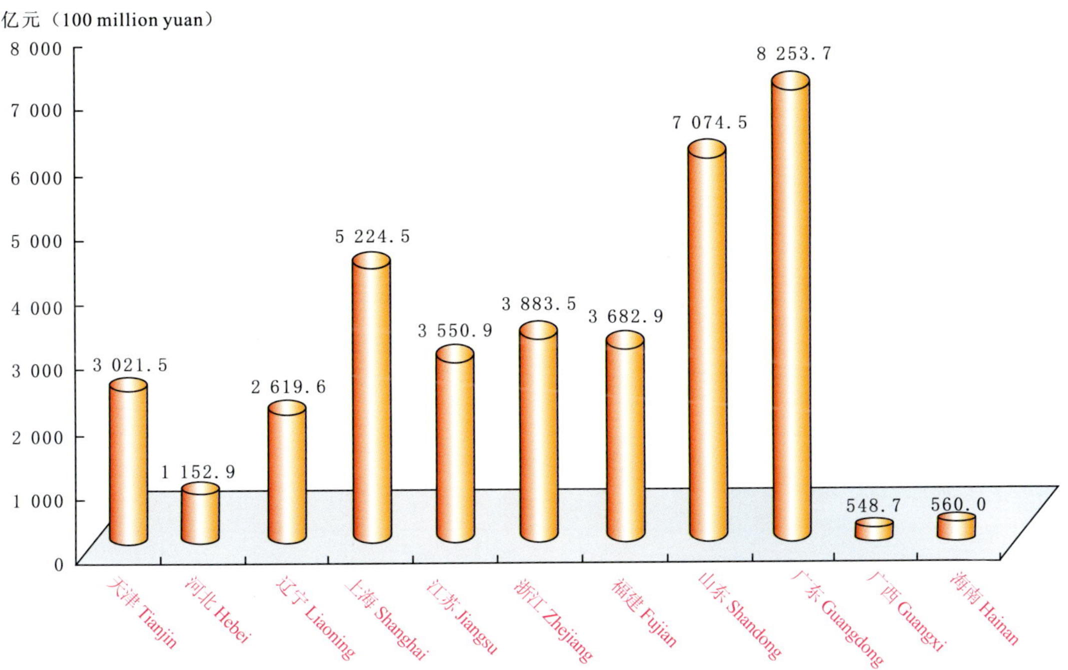

图4　全国海洋捕捞养殖产量
National Marine Catches and Mariculture Production

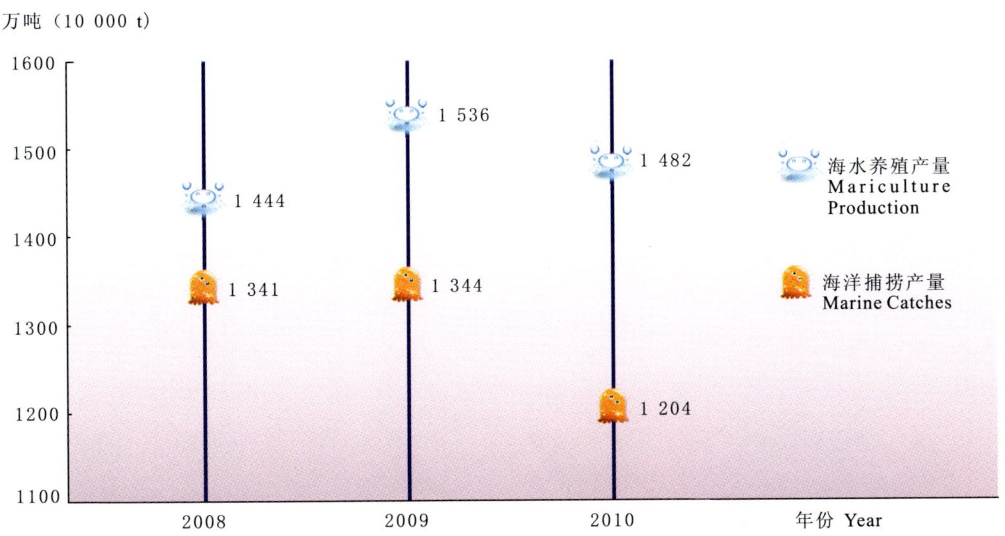

图5 2010年沿海地区海洋捕捞养殖产量
Marine Catches and Mariculture Production by Coastal Regions in 2010

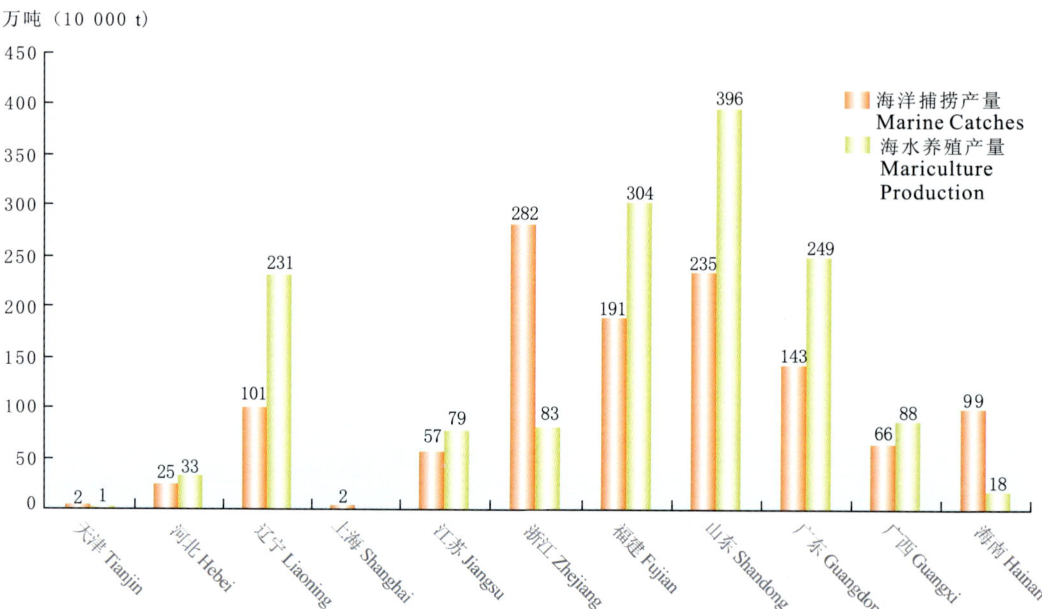

图6 全国海洋石油和天然气产量
National Output of Offshore Oil and Natural Gas

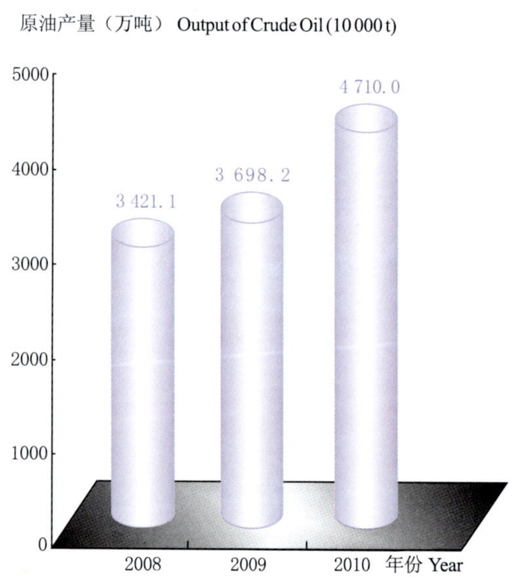

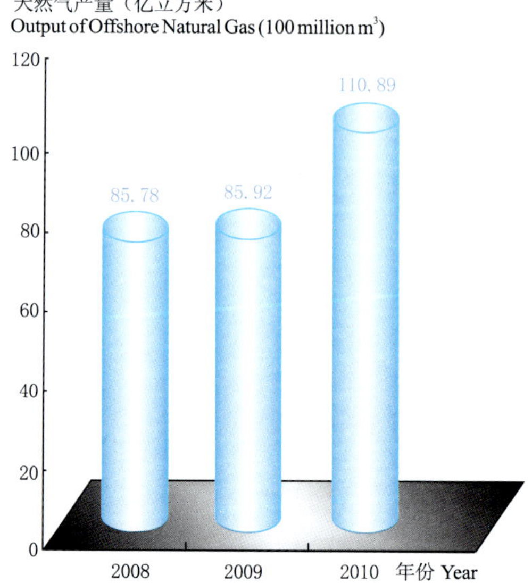

图7 海洋原油产量占全国原油产量比重

Proportion of Offshore Crude Oil Production in the National Total

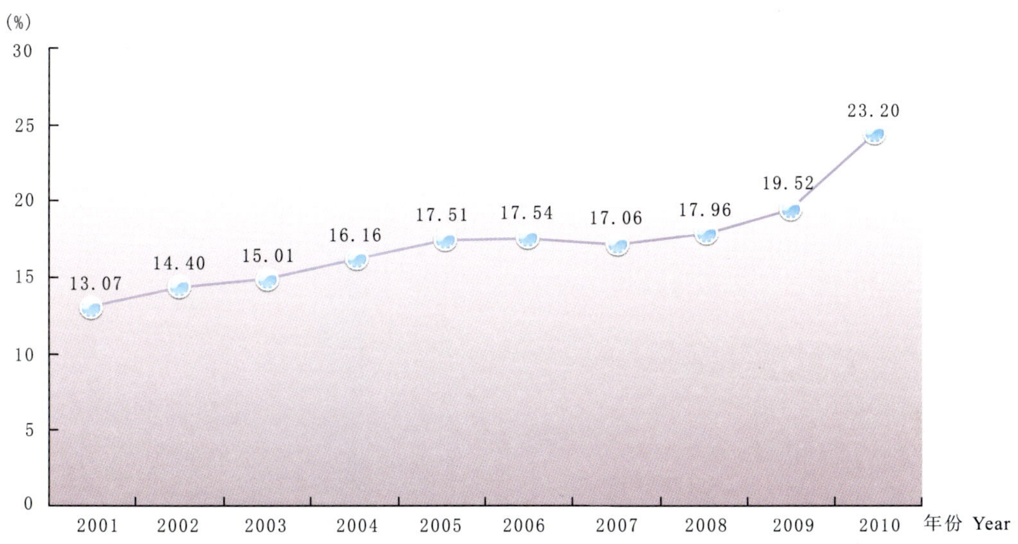

图8 2010年海洋矿业产量

Production of Marine Mining Industry in 2010

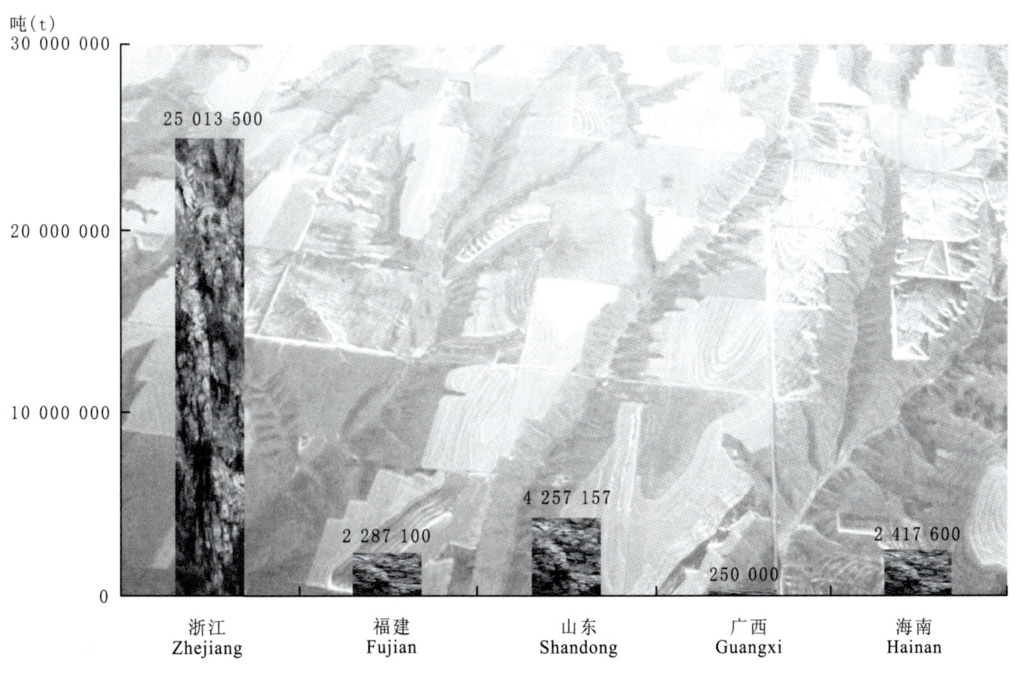

图9　2010年沿海地区海盐产量

Sea Salt Production by Coastal Regions in 2010

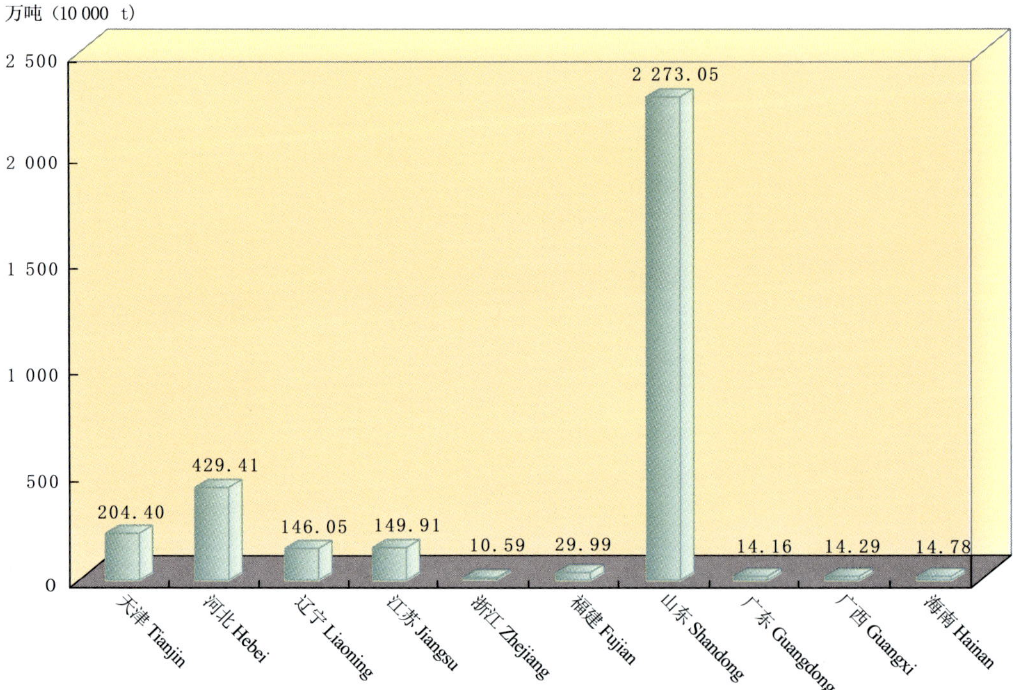

图10　2010年沿海地区造船完工量

Completed Number of Ships Built by Coastal Regions in 2010

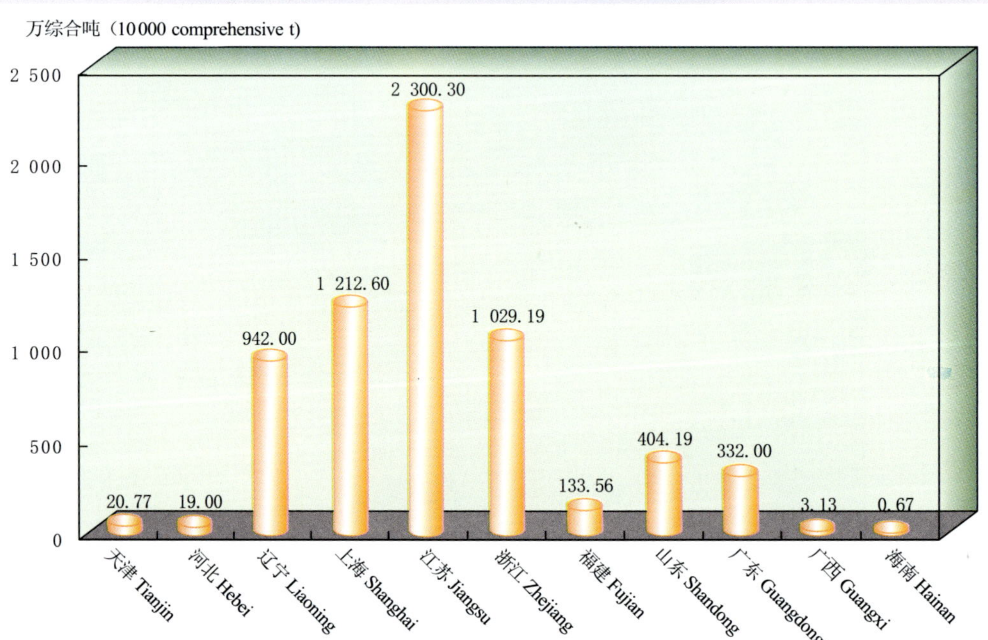

图11　2010年沿海地区海洋货物周转量
Goods Turnover Volume by Coastal Regions in 2010

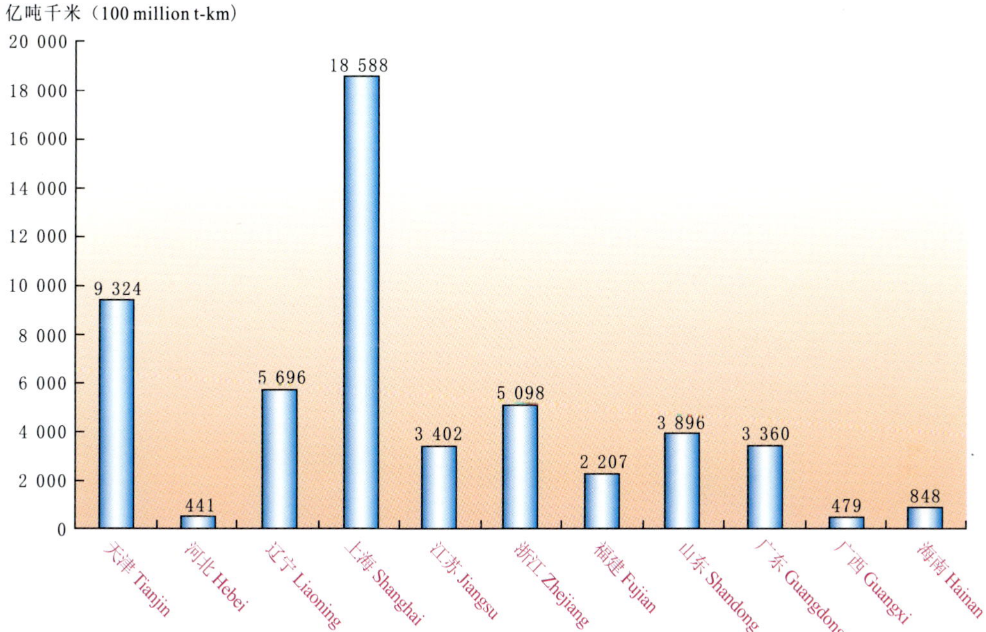

图12　2010年沿海港口国际标准集装箱吞吐量
International Standardized Containers Handled Coastal Seaports in 2010

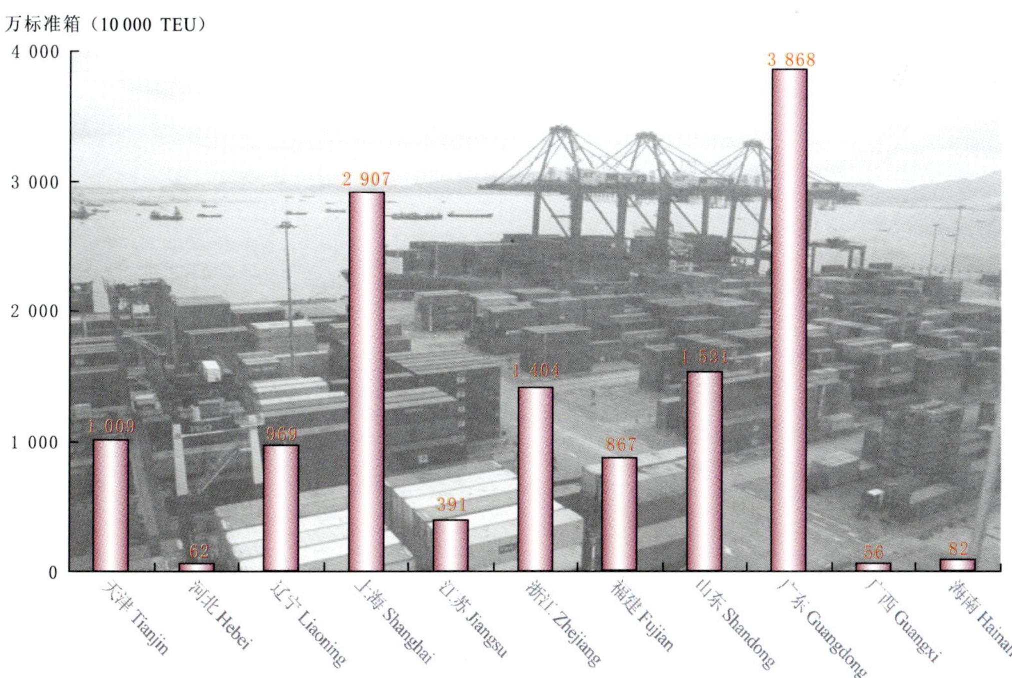

图13 主要沿海城市国际旅游（外汇）收入
Foreign Exchange Earnings from International Tourism by Coastal Cities

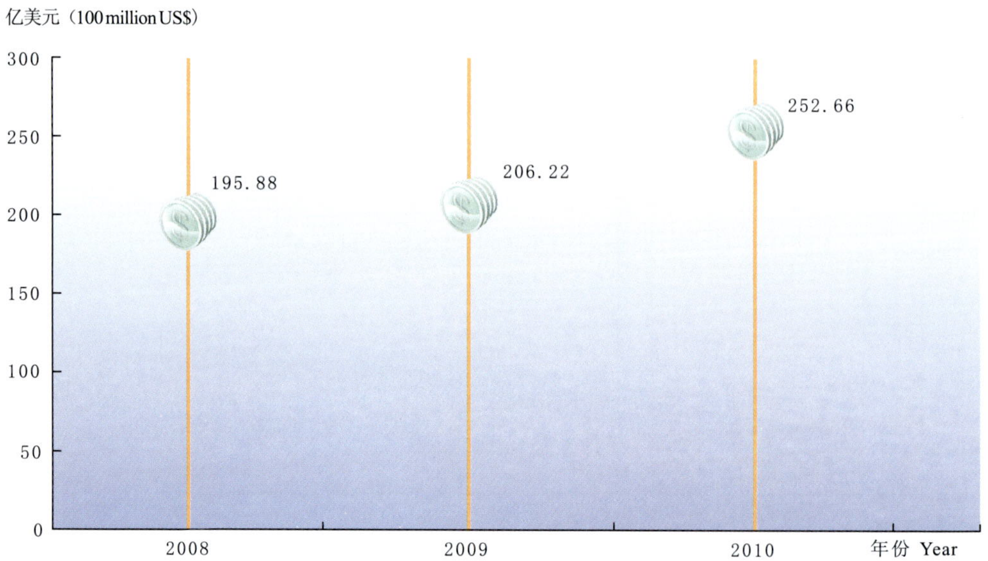

图14 主要沿海城市接待入境旅游者人数
Number of Oversea Visitor Arrivals in Major Coastal Cities

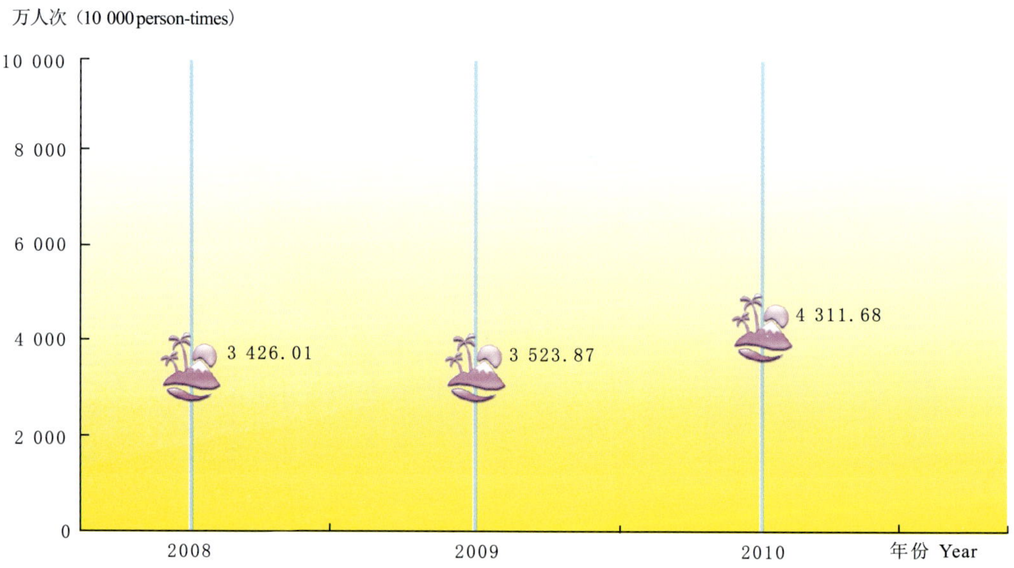

图15 2010年沿海地区海洋生产总值
Gross Ocean Product by Coastal Regions in 2010

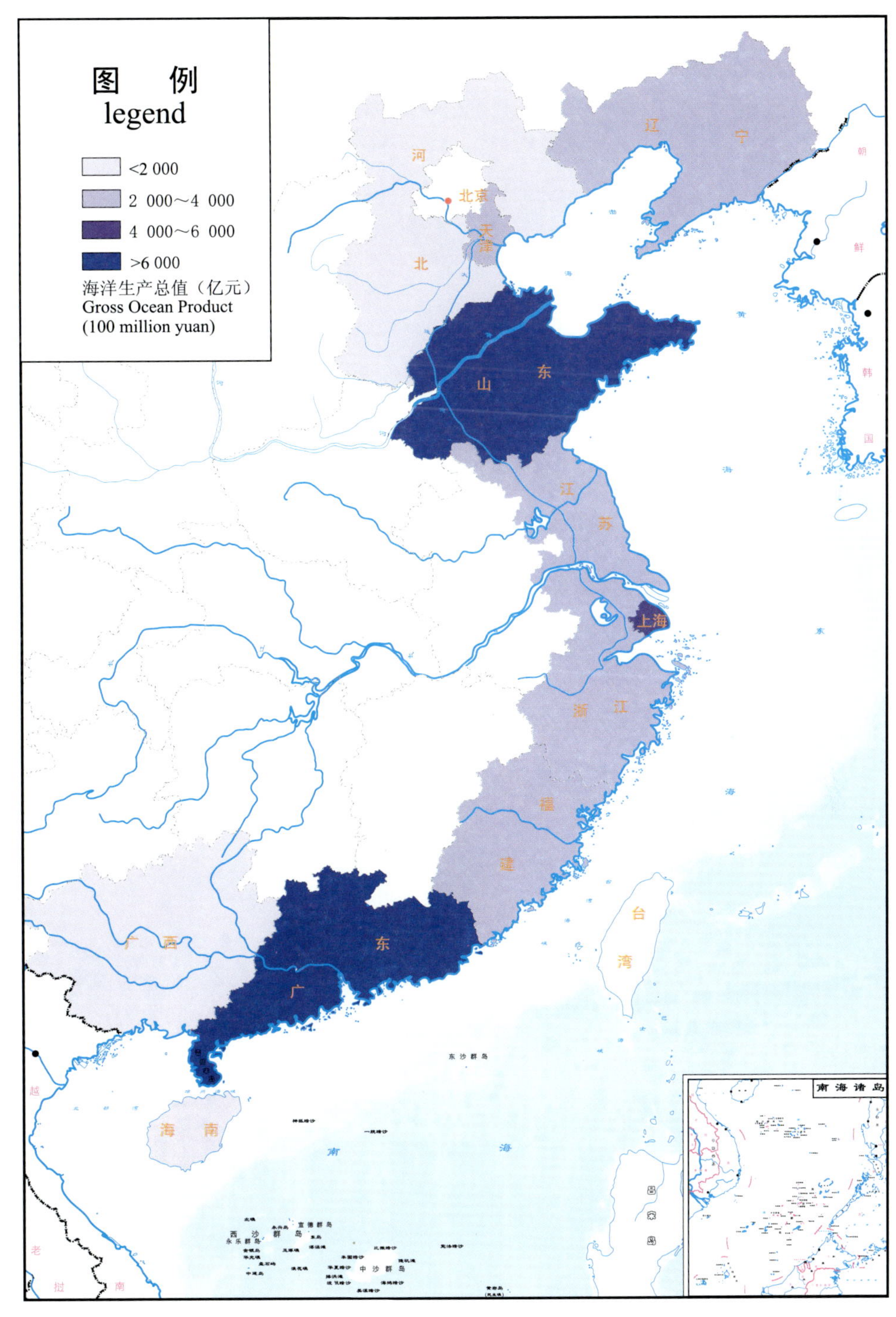

图16 2010年沿海地区海洋经济贡献

Marine Economic Contributions by Coastal Regions in 2010

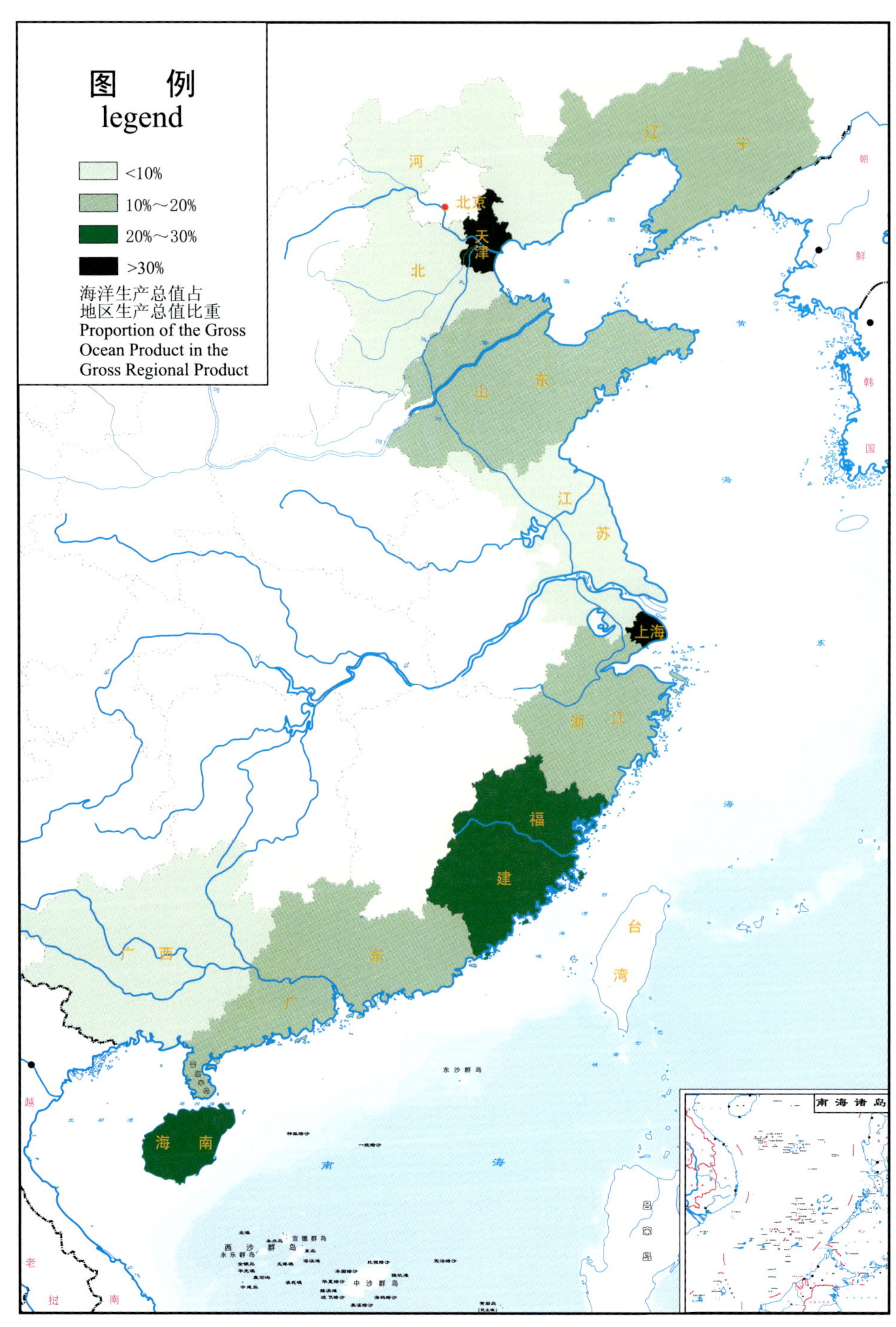

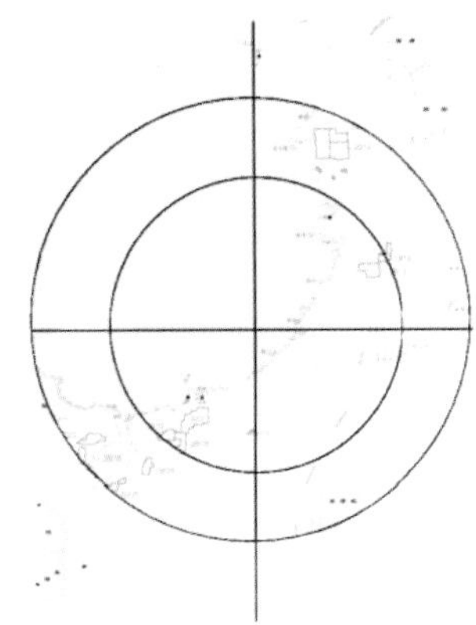

1

综 合 资 料

Integrated Data

1-1 沿海地区行政区划
Administrative Division of Coastal Regions

单位：个 (number)

沿海地区 Coastal Region	沿海城市 Coastal City	沿海地带 Coastal County (District)			
		合　计 Total	县 County	县级市 County-level city	区 District
合　计 Total	**53**	**237**	**64**	**56**	**117**
天　津 Tianjin	1	1			1
河　北 Hebei	3	11	6	1	4
辽　宁 Liaoning	6	22	4	7	11
上　海 Shanghai	1	5	1		4
江　苏 Jiangsu	3	15	8	4	3
浙　江 Zhejiang	7	35	11	10	14
福　建 Fujian	6	34	11	8	15
山　东 Shandong	7	37	6	14	17
广　东 Guangdong	14	56	11	6	39
广　西 Guangxi	3	8	1	1	6
海　南 Hainan	2	13	5	5	3

注：沿海地带中未包括广东省的东莞、中山和海南省的三亚。

Note: Dongguan, Zhongshan of Guangdong Province and Sanya of Hainan Province are not included in the Coastal County.

1-2 沿海行政区划一览表
Table of Administrative Division of Coastal Regions

沿海地区 Coastal Region	地区代码 Zip Code	沿海城市 Coastal City	地区代码 Zip Code	沿海地带 Coastal County (District)	地区代码 Zip Code
天 津 Tianjin	120000			滨海新区Binhai Xinqu	120116
河 北 Hebei	130000	唐山 Tangshan	130200	丰南区Fengnan Qu	130207
				滦南县Luannan Xian	130224
				乐亭县Leting Xian	130225
				唐海县Tanghai Xian	130230
		秦皇岛 Qinhuangdao	130300	海港区Haigang Qu	130302
				山海关区Shanhaiguan Qu	130303
				北戴河区Beidaihe Qu	130304
				昌黎县Changli Xian	130322
				抚宁县Funing Xian	130323
		沧州 Cangzhou	130900	海兴县Haixing Xian	130924
				黄骅市Huanghua Shi	130983
辽 宁 Liaoning	210000	大连 Dalian	210200	中山区Zhongshan Qu	210202
				西岗区Xigang Qu	210203
				沙河口区Shahekou Qu	210204
				甘井子区Ganjingzi Qu	210211
				旅顺口区Lüshunkou Qu	210212
				金州区Jinzhou Qu	210213
				长海县Changhai Xian	210224
				瓦房店市Wafangdian Shi	210281
				普兰店市Pulandian Shi	210282
				庄河市Zhuanghe Shi	210283
		丹东 Dandong	210600	东港市Donggang Shi	210681
		锦州 Jinzhou	210700	凌海市Linghai Shi	210781
		营口 Yingkou	210800	西市区Xishi Qu	210803
				鲅鱼圈区Bayuquan Qu	210804
				老边区Laobian Qu	210811
				盖州市Gaizhou Shi	210881
		盘锦 Panjin	211100	大洼县Dawa Xian	211121
				盘山县Panshan Xian	211122

1-2 续表1 continued

沿海地区 Coastal Region	地区代码 Zip Code	沿海城市 Coastal City	地区代码 Zip Code	沿海地带 Coastal County (District)	地区代码 Zip Code
		葫芦岛 Huludao	211400	连山区Lianshan Qu	211402
				龙港区Longgang Qu	211403
				绥中县Suizhong Xian	211421
				兴城市Xingcheng Shi	211481
上 海 Shanghai	310000			宝山区Baoshan Qu	310113
				浦东新区Pudong Xinqu	310115
				金山区Jinshan Qu	310116
				奉贤区Fengxian Qu	310120
				崇明县Chongming Xian	310230
江 苏 Jiangsu	320000	南通 Nantong	320600	通州区Tongzhou Qu	320612
				海安县Hai'an Xian	320621
				如东县Rudong Xian	320623
				启东市Qidong Shi	320681
				海门市Haimen Shi	320684
		连云港 Lianyungang	320700	连云区Lianyun Qu	320703
				新浦区Xinpu Qu	320705
				赣榆县Ganyu Xian	320721
				灌云县Guanyun Xian	320723
				灌南县Guannan Xian	320724
		盐城 Yancheng	320900	响水县Xiangshui Xian	320921
				滨海县Binhai Xian	320922
				射阳县Sheyang Xian	320924
				东台市Dongtai Shi	320981
				大丰市Dafeng Shi	320982
浙 江 Zhejiang	330000	杭州 Hangzhou	330100	滨江区Binjiang Qu	330108
				萧山区Xiaoshan Qu	330109
		宁波 Ningbo	330200	海曙区Haishu Qu	330203
				江东区Jiangdong Qu	330204
				江北区Jiangbei Qu	330205
				北仑区Beilun Qu	330206
				镇海区Zhenhai Qu	330211
				鄞州区Yinzhou Qu	330212
				象山县Xiangshan Xian	330225
				宁海县Ninghai Xian	330226
				余姚市Yuyao Shi	330281
				慈溪市Cixi Shi	330282
				奉化市Fenghua Shi	330283

1-2 续表2 continued

沿海地区 Coastal Region	地区代码 Zip Code	沿海城市 Coastal City	地区代码 Zip Code	沿海地带 Coastal County (District)	地区代码 Zip Code
		温州 Wenzhou	330300	龙湾区Longwan Qu	330303
				瓯海区Ouhai Qu	330304
				洞头县Dongtou Xian	330322
				平阳县Pingyang Xian	330326
				苍南县Cangnan Xian	330327
				瑞安市Rui'an Shi	330381
				乐清市Yueqing Shi	330382
		嘉兴 Jiaxing	330400	海盐县Haiyan Xian	330424
				海宁市Haining Shi	330481
				平湖市Pinghu Shi	330482
		绍兴 Shaoxing	330600	绍兴县Shaoxing Xian	330621
				上虞市Shangyu Shi	330682
		舟山 Zhoushan	330900	定海区Dinghai Qu	330902
				普陀区Putuo Qu	330903
				岱山县Daishan Xian	330921
				嵊泗县Shengsi Xian	330922
		台州 Taizhou	331000	椒江区Jiaojiang Qu	331002
				路桥区Luqiao Qu	331004
				玉环县Yuhuan Xian	331021
				三门县Sanmen Xian	331022
				温岭市Wenling Shi	331081
				临海市Linhai Shi	331082
福建 Fujian	350000	福州 Fuzhou	350100	马尾区Mawei Qu	350105
				连江县Lianjiang Xian	350122
				罗源县Luoyuan Xian	350123
				平潭县Pingtan Xian	350128
				福清市Fuqing Shi	350181
				长乐市Changle Shi	350182
		厦门 Xiamen	350200	思明区Siming Qu	350203
				海沧区Haicang Qu	350205

1-2 续表3 continued

沿海地区 Coastal Region	地区代码 Zip Code	沿海城市 Coastal City	地区代码 Zip Code	沿海地带 Coastal County (District)	地区代码 Zip Code
				湖里区Huli Qu	350206
				集美区Jimei Qu	350211
				同安区Tong'an Qu	350212
				翔安区Xiang'an Qu	350213
		莆田 Putian	350300	城厢区Chengxiang Qu	350302
				涵江区Hanjiang Qu	350303
				荔城区Licheng Qu	350304
				秀屿区Xiuyu Qu	350305
				仙游县Xianyou Xian	350322
		泉州 Quanzhou	350500	丰泽区Fengze Qu	350503
				洛江区Luojiang Qu	350504
				泉港区Quangang Qu	350505
				惠安县Hui'an Xian	350521
				金门县Jinmen Xian	350527
				石狮市Shishi Shi	350581
				晋江市Jinjiang Shi	350582
				南安市Nan'an Shi	350583
		漳州 Zhangzhou	350600	云霄县Yunxiao Xian	350622
				漳浦县Zhangpu Xian	350623
				诏安县Zhao'an Xian	350624
				东山县Dongshan Xian	350626
				龙海市Longhai Shi	350681
		宁德 Ningde	350900	蕉城区Jiaocheng Qu	350902
				霞浦县Xiapu Xian	350921
				福安市Fu'an Shi	350981
				福鼎市Fuding Shi	350982
山东 Shandong	370000	青岛 Qingdao	370200	市南区Shinan Qu	370202
				市北区Shibei Qu	370203
				四方区Sifang Qu	370205
				黄岛区Huangdao Qu	370211
				崂山区Laoshan Qu	370212
				李沧区Licang Qu	370213
				城阳区Chengyang Qu	370214
				胶州市Jiaozhou Shi	370281
				即墨市Jimo Shi	370282
				胶南市Jiaonan Shi	370284

1-2 续表4 continued

沿海地区 Coastal Region	地区代码 Zip Code	沿海城市 Coastal City	地区代码 Zip Code	沿海地带 Coastal County (District)	地区代码 Zip Code
		东营 Dongying	370500	东营区Dongying Qu	370502
				河口区Hekou Qu	370503
				垦利县Kenli Xian	370521
				利津县Lijin Xian	370522
				广饶县Guangrao Xian	370523
		烟台 Yantai	370600	芝罘区Zhifu Qu	370602
				福山区Fushan Qu	370611
				牟平区Muping Qu	370612
				莱山区Laishan Qu	370613
				长岛县Changdao Xian	370634
				龙口市Longkou Shi	370681
				莱阳市Laiyang Shi	370682
				莱州市Laizhou Shi	370683
				蓬莱市Penglai Shi	370684
				招远市Zhaoyuan Shi	370685
				海阳市Haiyang Shi	370687
		潍坊 Weifang	370700	寒亭区Hanting Qu	370703
				寿光市Shouguang Shi	370783
				昌邑市Changyi Shi	370786
		威海 Weihai	371000	环翠区Huancui Qu	371002
				文登市Wendeng Shi	371081
				荣成市Rongcheng Shi	371082
				乳山市Rushan Shi	371083
		日照 Rizhao	371100	东港区Donggang Qu	371102
				岚山区Lanshan Qu	371103
		滨州 Binzhou	371600	无棣县Wudi Xian	371623
				沾化县Zhanhua Xian	371624
广东 Guangdong	440000	广州 Guangzhou	440100	荔湾区Liwan Qu	440103
				越秀区Yuexiu Qu	440104
				海珠区Haizhu Qu	440105
				天河区Tianhe Qu	440106
				白云区Baiyun Qu	440111
				黄埔区Huangpu Qu	440112
				番禺区Panyu Qu	440113
				南沙区Nansha Qu	440115
				萝岗区Luogang Qu	440116

1-2 续表5 continued

沿海地区 Coastal Region	地区代码 Zip Code	沿海城市 Coastal City	地区代码 Zip Code	沿海地带 Coastal County (District)	地区代码 Zip Code
		深圳 Shenzhen	440300	罗湖区Luohu Qu	440303
				福田区Futian Qu	440304
				南山区Nanshan Qu	440305
				宝安区Bao'an Qu	440306
				龙岗区Longgang Qu	440307
				盐田区Yantian Qu	440308
		珠海 Zhuhai	440400	香洲区Xiangzhou Qu	440402
				斗门区Doumen Qu	440403
				金湾区Jinwan Qu	440404
		汕头 Shantou	440500	龙湖区Longhu Qu	440507
				金平区Jinping Qu	440511
				濠江区Haojiang Qu	440512
				潮阳区Chaoyang Qu	440513
				潮南区Chaonan Qu	440514
				澄海区Chenghai Qu	440583
				南澳县Nan'ao Xian	440523
		江门 Jiangmen	440700	蓬江区Pengjiang Qu	440703
				江海区Jianghai Qu	440704
				新会区Xinhui Qu	440705
				台山市Taishan Shi	440781
				恩平市Enping Shi	440785
		湛江 Zhanjiang	440800	赤坎区Chikan Qu	440802
				霞山区Xiashan Qu	440803
				坡头区Potou Qu	440804
				麻章区Mazhang Qu	440811
				遂溪县Suixi Xian	440823
				徐闻县Xuwen Xian	440825
				廉江市Lianjiang Shi	440881
				雷州市Leizhou Shi	440082
				吴川市Wuchuan Shi	440083
		茂名 Maoming	440900	茂南区Maonan Qu	440902
				茂港区Maogang Qu	440903
				电白县Dianbai Xian	440923
		惠州 Huizhou	441300	惠城区Huicheng Qu	441302
				惠阳区Huiyang Qu	441303
				惠东县Huidong Xian	441323

1-2 续表6 continued

沿海地区 Coastal Region	地区代码 Zip Code	沿海城市 Coastal City	地区代码 Zip Code	沿海地带 Coastal County (District)	地区代码 Zip Code
		汕尾 Shanwei	441500	城 区Chengqu	441502
				海丰县Haifeng Xian	441521
				陆丰市Lufeng Shi	441581
		阳江 Yangjiang	441700	江城区Jiangcheng Qu	441702
				阳西县Yangxi Xian	441721
				阳东县Yangdong Xian	441723
		东莞 Dongguan	441900		
		中山 Zhongshan	442000		
		潮州 Chaozhou	445100	湘桥区Xiangqiao Qu	445102
				饶平县Raoping Xian	445122
		揭阳 Jieyang	445200	榕城区Rongcheng Qu	445202
				揭东县Jiedong Xian	445221
				惠来县Huilai Xian	445224
广西 Guangxi	450000	北海 Beihai	450500	海城区Haicheng Qu	450502
				银海区Yinhai Qu	450503
				铁山港区Tieshangang Qu	450512
				合浦县Hepu Xian	450521
		防城港 Fangchenggang	450600	港口区Gangkou Qu	450602
				防城区Fangcheng Qu	450603
				东兴市Dongxing Shi	450681
		钦州 Qinzhou	450700	钦南区Qinnan Qu	450702
海南 Hainan	460000	海口 Haikou	460100	秀英区Xiuying Qu	460105
				龙华区Longhua Qu	460106
				美兰区Meilan Qu	460108
		三亚 Sanya	460200		
				琼海市 Qionghai Shi	469002
				儋州市 Danzhou Shi	469003
				文昌市 Wenchang Shi	469005
				万宁市 Wanning Shi	469006
				东方市 Dongfang Shi	469007
				澄迈县Chengmai Xian	469023
				临高县Lingao Xian	469024
				昌江黎族自治县 Changjiang Lizu Zizhixian	469026
				乐东黎族自治县 Ledong Lizu Zizhixian	469027
				陵水黎族自治县 Lingshui Lizu Zizhixian	469028

1-3 海洋自然地理
Marine Physical Geography

指 标		Item	指标值 Data
海洋平均深度	（米）	Average Depth of Sea (m)	961
海洋最大深度	（米）	Maximum Depth of Sea (m)	5 377
岸线总长度	（千米）	Length of Coastline (km)	32 000
大陆岸线长度		Mainland Shore	18 000
岛屿岸线长度		Island Shore	14 000
岛屿个数	（个）	Number of Islands (unit)	5 400
岛屿面积	（万平方千米）	Area of Islands (10 000 km^2)	3.87

注：数据来源于《2011 中国统计年鉴》（表1-4同）。岛屿面积未包括香港、澳门特别行政区和台湾省。

Note: The data come from the *China Statistical Yearbook 2010*（The same as in the Table 1-4）. Island area does not include that of Hong Kong Special Administrative Region, Macao Specical Administrative Region and Taiwan Province.

1-4 海区海域面积
Sea Area

自然海区名称	Natural Sea Area	海域总面积（千公顷）Sea Area (1 000 hm^2)	平均深度（米）Average Depth (m)	最大深度（米）Maximum Depth (m)
合 计	**Total**	**472 700**		
渤 海	Bohai Sea	7 700	18	70
黄 海	Huanghai Sea	38 000	44	140
东 海	Donghai Sea	77 000	370	2 719
南 海	Nanhai Sea	350 000	1 212	5 559

1-5 海洋自然资源
Ocean Natural Resources

指　　标	Item	指标值 Data
海洋能源理论蕴藏量　（亿千瓦）	Theoretical Sea-energy Reserves (100 million kW)	6.3
海水可养殖面积　（万公顷）	Cultivatable Area in Marine Areas (10 000 hm^2)	260.01
浅海滩涂可养殖面积　（万公顷）	Cultivable Area in Shallow Sea and Sea-beaches (10 000 hm^2)	242.00
大陆架渔场面积　（万公顷）	Area of Fishing Ground of Continental Shelf (10 000 hm^2)	28 000.00

注：数据来源于《2011中国统计年鉴》。

Note: The data come from the *China Statistical Yearbook 2011*.

1-6 海区海洋石油储量
Offshore Oil Reserves in the Sea Area

自然海区名称 Natural Sea Area	海洋石油（万吨） Offshore Oil (10 000 t)	
	累计探明技术可采储量 Proven Technically Recoverable Reserves in the Aggregate	剩余技术可采储量 Surplus Technically Recoverable Reserves
合 计 Total	**82 554.9**	**44 012.5**
渤 海 Bohai Sea	47 134.8	31 305.2
黄 海 Huanghai Sea		
东 海 Donghai Sea	1 221.7	826.6
南 海 Nanhai Sea	34 198.4	11 880.7

注：数据来源于《2010年全国矿产资源储量通报》。

Note: The data come from the *Journal on the National Mineral Resources Reserves in 2010*.

1-7 沿海地区水资源基本情况
Water Resources by Coastal Regions

地 区 Region	水资源总量 （亿立方米） Total Amount of Water Resources (100 million m³)	地表 水资源量 Surface Water Resources	地下 水资源量 Groundwater Resources	地表水与地下 水资源重复量 Duplicated Measurement Between Surface Water and Groundwater	人均水资源量 （立方米/人） Per Capita Water Resources (m³/person)
全国总计 National Total	**30 906.4**	**29 797.6**	**8 417.0**	**7 308.2**	**2 310.4**
天 津 Tianjin	9.2	5.6	4.5	0.8	72.8
河 北 Hebei	138.9	56.6	112.9	30.6	195.3
辽 宁 Liaoning	606.7	554.0	146.8	94.1	1 392.1
上 海 Shanghai	36.8	30.9	8.9	3.0	163.1
江 苏 Jiangsu	383.5	291.2	108.9	16.6	489.2
浙 江 Zhejiang	1 398.6	1 382.9	264.7	249.1	2 608.7
福 建 Fujian	1 652.7	1 651.5	353.8	352.6	4 491.7
山 东 Shandong	309.1	199.1	181.2	71.2	324.4
广 东 Guangdong	1 998.8	1 989.5	478.3	468.9	1 943.3
广 西 Guangxi	1 823.6	1 823.6	355.8	355.8	3 852.9
海 南 Hainan	479.8	474.3	105.7	100.2	5 538.7

注：数据来源于《2011中国统计年鉴》。

Note: The data come from the *China Statistical Yearbook 2011*.

1-8 沿海地区湿地面积
Area of Wetlands by Coastal Regions

地 区 Region	湿地面积 （千公顷） Area of Wetlands (1 000 hm^2)	近岸及海岸 Coasts and Seashores	湿地面积 占国土面积 比重 （%） Proportion of Wetlands in Total Area of Territory (%)
全国总计 National Total	**38 485.5**	**5 941.7**	**4.01**
天 津 Tianjin	171.8	58.1	14.95
河 北 Hebei	1 081.9	278.8	5.82
辽 宁 Liaoning	1 219.6	738.1	8.37
上 海 Shanghai	319.7	305.4	53.68
江 苏 Jiangsu	1 674.7	843.5	16.32
浙 江 Zhejiang	802.2	574.3	7.88
福 建 Fujian	443.0	370.6	3.65
山 东 Shandong	1 784.1	1 210.9	11.72
广 东 Guangdong	1 398.1	1 017.8	7.86
广 西 Guangxi	656.1	348.4	2.76
海 南 Hainan	311.5	190.0	9.13

注：数据来源于国家林业局。本表为中国首次湿地调查 (1995～2003)资料，不包括台湾省、香港和澳门特别行政区；湿地面积不包括水稻田湿地。

Note: The data come from the State Forestry Administration. Data in the table are the figures of China First Wetlands Survey (1995-2003), excluding the wetlands of Taiwan province, Hong Kong Special Administrative Region and Macao SAR. Area of wetlands excludes the wetland of paddyfield.

1-9 红树林各地类面积
Site Classification and Area of Sharpleaf Mangrove (Rhizophora Apiculata)

单位：公顷　　　　(hm^2)

地　区 Region	红树林各地类总面积 Total Site Area of Sharpleaf Mangrove	现有面积 Established	未成林面积 Unestablished	宜林地面积 Suitable for Planting
全国总计 **National Total**	**82 757.2**	**22 024.9**	**1 884.1**	**58 848.2**
浙　江 Zhejiang	5 452.3	20.6	236.1	5 195.6
福　建 Fujian	13 410.1	615.1	286.4	12 508.6
广　东 Guangdong	32 325.9	9 084.0	981.3	22 260.6
广　西 Guangxi	18 029.2	8 374.9	380.3	9 274.0
海　南 Hainan	13 539.7	3 930.3		9 609.4

注：数据来源于《2011中国统计年鉴》。本表数据为2002年全国红树林资源调查资料。

Note: The data come from the *China Statistical Yearbook 2011* . Data in the table are the figures of National Sharpleaf Mangrove Survey in 2002.

1-10 主要沿海城市气候基本情况
Climate of Major Coastal Cities

城 市 City	年平均气温 （摄氏度） Annual Average Temperature(℃)	年平均相对湿度 （%） Annual Average Relative Humidity (%)	全年降水量 （毫米） Annual Precipitation (mm)	全年日照时数 （小时） Annual Sunshine Hours (h)
天 津 Tianjin	12.2	59	355.4	2 089.2
上 海 Shanghai	17.2	69	1 128.9	1 662.0
杭 州 Hangzhou	17.4	72	1 728.1	1 689.3
福 州 Fuzhou	20.4	74	1 604.5	1 485.6
广 州 Guangzhou	22.5	73	2 353.6	1 484.0
海 口 Haikou	24.6	81	2 445.1	1 823.3

注：数据来源于《2011中国统计年鉴》。

Note: The data come from the *China Statistical Yearbook 2011*.

主要统计指标解释

1. 沿海地区 即广义的沿海地区是指有海岸线（大陆岸线和岛屿岸线）的地区，按行政区划分为沿海省、自治区、直辖市。

2. 沿海城市 是指有海岸线的直辖市和地级市（包括其下属的全部区、县和县级市）。

3. 沿海地带 即狭义的沿海地区，是指有海岸线的县、县级市、区（包括直辖市和地级市的区）。

4. 海洋 是海和洋的统称。洋为地球表面上相连接的广大咸水水体的主体部分。海为地球表面相连接的广大咸水水体被陆地、岛礁、半岛包围或分隔的边缘部分。

5. 海水可养殖面积 指利用滩涂、浅海、港湾进行鱼、虾、蟹、贝、藻等海水经济动植物的人工养殖的水面面积。

6. 水资源总量 指评价区内降水形成的地表和地下产水总量，即地表产流量与降水入渗补给地下水量之和，不包括过境水量。

7. 地表水资源量 指评价区内河流、湖泊、冰川等地表水体中可以逐年更新的动态水量，即当地天然河川径流量。

8. 地下水资源量 指评价区内降水和地表水对饱水岩土层的补给量，包括降水入渗补给量和河道、湖库、渠系、渠灌田间等地表水体的入渗补给量。

9. 地表水与地下水资源重复量 指地表水和地下水相互转化的部分，即天然河川径流量中的地下水排泄量和地下水补给量中来源于地表水的入渗补给量。

10. 湿地 指天然或人工、长久或暂时性的沼泽地、泥炭地或水域地带，包括静止或流动、淡水、半咸水、咸水体，低潮时水深不超过6米的水域以及海岸地带地区的珊瑚滩和海草床、滩涂、红树林、河口、河流、淡水沼泽、沼泽森林、湖泊、盐沼及盐湖。

11. 红树林 指生长在热带、亚热带低能海岸潮间带上部，受周期性潮水浸淹，以红树植物为主体的常绿灌木或乔木组成的潮滩湿地木本生物群落。

12. 气温 指空气的温度，我国一般以摄氏度(℃)为单位表示。气象观测的温度表是放在离地面约1.5米处通风良好的百叶箱里测量的，因此，通常说的气温指的是离地面1.5米处百叶箱中的温度。其统计计算方法为：

月平均气温是将全月各日的平均气温相加，除以该月的天数而得。

年平均气温是将12个月的月平均气温累加后除以12而得。

13. 相对湿度 指空气中实际所含水蒸气密度和同温度下饱和水蒸气密度的百分比值。其统计方法与气温相同。

14. 降水量 指从天空降落到地面的液态或固态(经融化后)水，未经蒸发、渗透、流失而在地面上积聚的深度。其统计计算方法为：

月降水量是将全月各日的降水量累加而得。

年降水量是将12个月的月降水量累加而得。

15. 日照时数 指太阳实际照射地面的时间。其统计方法与降水量相同。

Explanatory Notes on Main Statistical Indicators

1. Coastal Region, i.e., the coastal region in a broad sense, refers to the regions with coastlines (continental and island coastlines), which are divided into the coastal provinces, autonomous regions and municipalities directly under the Central Government according to the administrative zoning.

2. Coastal City refers to the municipalities directly under the Central Government and the prefecture-level cities (including all the districts, counties and county-level cities under them).

3. Coastal Zone, i.e., the coastal region in a narrow sense, refers to the counties, county-level cities and districts with coastlines (including the districts under the municipalities directly under the Central Government and the prefecture-level districts).

4. Ocean is the general name for sea and ocean. Ocean refers to the main body of large salt water connected with the earth surface . Sea refers to the edge areas of the salt water on the earth surface that are compartmentalized or surrounded by land, island, reef or peninsula.

5. Marine Cultivatable Areas refer to water areas in beach, shallow sea and bays that are used to breed marine cash propagation, such as fish, shrimp, crab, shellfish, alga and so on.

6. Total Water Resources refers to total volume of water resources measured as run-off for surface water from rainfall and recharge for groundwater in a given area, excluding transit water.

7. Surface Water Resources refers to total renewable resources which exist in rivers, lakes, glaciers and other collectors from rainfall and are measured as run-off of rivers.

8. Groundwater Resources refers to replenishment of aquifers with rainfall and surface water.

9. Duplicated Measurement between Surface Water and Groundwater refers to the exchange between surface water and groundwater, i.e. run-off of rivers includes some depletion into groundwater while groundwater includes some replenishment from surface water.

10. Wetlands refer to marshland and peat bog, whether natural or man-made, permanent or temporary; water covered areas, whether stagnant or flowing, with fresh or brackish-fresh or salty water that is less than 6 meters deep at low tide; as well as coral beach, weed beach, mud beach, mangrove, river outlet, rivers, fresh-water marshland, marshland forests, lakes, salty bog and salt lakes along the coastal areas.

11. Mangrove refers to evergreen woody plants or plant communities in tropical or sub-tropical zones which live between the sea and the land in areas which are inundated by tides.

12. Temperature refers to the air temperature. China uses centigrade as the unit. The thermometry used for weather observation is put in a breezy shutter, which is 1.5 meters high from the ground. Therefore, the commonly used temperature refers to the temperature in the breezy shutter 1.5 meters away from the ground. The calculation method is as follows:

Monthly average temperature is the summation of average daily temperature of one month divided by the actual days of that particular month.

Annual average temperature is the summation of monthly averages of a year divided by 12 months.

13. Relative Humidity refers to the ratio of actual water vapour pressure to the saturated water vapour density under the current temperature. The calculation method is the same as that of temperature.

14. Volume of Precipitation refers to the deepness of liquid state or solid state (thawed) water falling from the sky to the ground that has not evaporated, infiltrated or run off. The calculation method is as follows:

Monthly precipitation is the summation of daily precipitation of a month.

Annual precipitation is the summation of 12 months precipitation of a year.

15. Sunshine Hours refer to the actual hours of sun irradiating the earth. The calculation method is the same as that of the precipitation.

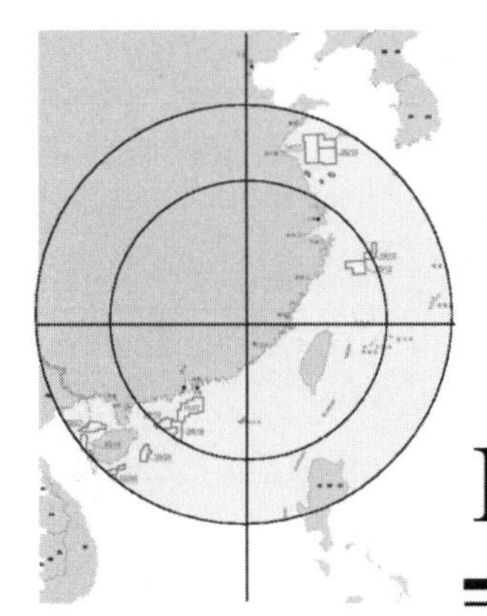

2

海洋经济核算

Marine Economic Accounting

2-1　全国海洋生产总值
National Gross Ocean Product

年　份 Year	海洋生产总值（亿元） Gross Ocean Product (100 million yuan)	第一产业 Primary Industry	第二产业 Secondary Industry	第三产业 Tertiary Industry	海洋生产总值占国内生产总值比重（%） Proportion of the Gross Ocean Product in GDP (%)	海洋生产总值增长速度（%） Growth Rate of the Gross Ocean Product (%)
2001	9 518.4	646.3	4 152.1	4 720.1	8.68	
2002	11 270.5	730.0	4 866.2	5 674.3	9.37	19.8
2003	11 952.3	766.2	5 367.6	5 818.5	8.80	4.2
2004	14 662.0	851.0	6 662.8	7 148.2	9.17	16.9
2005	17 655.6	1 008.9	8 046.9	8 599.8	9.55	16.3
2006	21 592.4	1 228.8	10 217.8	10 145.7	9.98	18.0
2007	25 618.7	1 395.4	12 011.0	12 212.3	9.64	14.8
2008	29 718.0	1 694.3	13 735.3	14 288.4	9.46	9.9
2009	32 277.6	1 857.7	14 980.3	15 439.5	9.47	9.2
2010	39 572.7	2 008.0	18 935.0	18 629.8	9.86	14.7

2-2　全国海洋生产总值构成
Composition of National Gross Ocean Product

年　份 Year	第一产业 Primary Industry (%)	第二产业 Secondary Industry (%)	第三产业 Tertiary Industry (%)
2001	6.8	43.6	49.6
2002	6.5	43.2	50.3
2003	6.4	44.9	48.7
2004	5.8	45.4	48.8
2005	5.7	45.6	48.7
2006	5.7	47.3	47.0
2007	5.4	46.9	47.7
2008	5.7	46.2	48.1
2009	5.8	46.4	47.8
2010	5.1	47.8	47.1

2-3　海洋及相关产业增加值
Added Values of Marine and Related Industries

单位：亿元　　(100 million yuan)

年　份 Year	合　计 Total	海洋产业 Marine Industry	主要海洋产业 Major Marine Industry	海洋科研教育管理服务业 Industries of Marine Scientific Research, Education, Management and Service	海洋相关产业 Ocean-related Industries
2001	9 518.4	5 733.6	3 856.6	1 877.0	3 784.8
2002	11 270.5	6 787.3	4 696.8	2 090.5	4 483.2
2003	11 952.3	7 137.7	4 754.4	2 383.3	4 814.6
2004	14 662.0	8 710.1	5 827.7	2 882.5	5 951.9
2005	17 655.6	10 539.0	7 188.0	3 350.9	7 116.6
2006	21 592.4	12 696.7	8 790.4	3 906.4	8 895.6
2007	25 618.7	15 070.6	10 478.3	4 592.3	10 548.0
2008	29 718.0	17 591.2	12 176.0	5 415.2	12 126.8
2009	32 277.6	18 822.0	12 843.6	5 978.4	13 455.6
2010	39 572.7	22 831.0	16 187.8	6 643.1	16 741.7

2-4　海洋及相关产业增加值构成
Composition of the Added Value of Marine and Related Industries

单位：%　　(%)

年　份 Year	合　计 Total	海洋产业 Marine Industry	主要海洋产业 Major Marine Industry	海洋科研教育管理服务业 Industries of Marine Scientific Research, Education, Management and Service	海洋相关产业 Ocean-related Industries
2001	100.0	60.2	40.5	19.7	39.8
2002	100.0	60.2	41.7	18.5	39.8
2003	100.0	59.7	39.8	19.9	40.3
2004	100.0	59.4	39.7	19.7	40.6
2005	100.0	59.7	40.7	19.0	40.3
2006	100.0	58.8	40.7	18.1	41.2
2007	100.0	58.8	40.9	17.9	41.2
2008	100.0	59.2	41.0	18.2	40.8
2009	100.0	58.3	39.8	18.5	41.7
2010	100.0	57.7	40.9	16.8	42.3

2-5 全国主要海洋产业增加值
Gross Output Value and Added Value of Major Marine Industries

主要海洋产业 Major Marine Industry	增加值 （亿元） Added Value (100 million yuan)	比上年增长（%） (按可比价计算) Percentage of Increase Over Last Year (%) (at comparable price)
合计 **Total**	**16 187.8**	**17.4**
海洋渔业 Marine Fishery Industry	2 851.6	6.0
海洋油气业 Offshore Oil and Natural Gas Industry	1 302.2	53.9
海洋矿业 Marine Mining Industry	45.2	- 8.8
海洋盐业 Sea Salt Industry	65.5	13.7
海洋船舶工业 Marine Shipbuilding Industry	1 215.6	22.8
海洋化工业 Marine Chemical Industry	613.8	22.1
海洋生物医药业 Marine Biomedicine Industry	83.8	29.9
海洋工程建筑业 Marine Engineering Architecture Industry	874.2	23.9
海洋电力业 Marine Electric Power Industry	38.1	80.1
海水利用业 Marine Seawater Utilization Industry	8.9	9.0
海洋交通运输业 Maritime Communications and Transportation Industry	3 785.8	20.6
滨海旅游业 Coastal Tourism	5 303.1	13.3

2-6 海洋渔业增加值
Added Value of Marine Fishery Industry

单位：亿元 (100 million yuan)

年 份 Year	增加值 Added Value
2001	966.0
2002	1 091.2
2003	1 145.0
2004	1 271.2
2005	1 507.6
2006	1 672.0
2007	1 906.0
2008	2 228.6
2009	2 440.8
2010	2 851.6

2-7 海洋油气业增加值
Added Value of Offshore Oil and Gas Industry

单位：亿元 (100 million yuan)

年 份 Year	增加值 Added Value
2001	176.8
2002	181.8
2003	257.0
2004	345.1
2005	528.2
2006	668.9
2007	666.9
2008	1 020.5
2009	614.1
2010	1 302.2

2-8 海洋矿业增加值
Added Value of Marine Mining Industry

单位：亿元 (100 million yuan)

年 份 Year	增加值 Added Value
2001	1.0
2002	1.9
2003	3.1
2004	7.9
2005	8.3
2006	13.4
2007	16.3
2008	35.2
2009	41.6
2010	45.2

注：自2008年起部分地区统计矿种增加。

Note: The data from 2008 include added kinds of minerals in some regions.

2-9 海洋盐业增加值
Added Value of Marine Salt Industry

单位：亿元 (100 million yuan)

年 份 Year	增加值 Added Value
2001	32.6
2002	34.2
2003	28.4
2004	39.0
2005	39.1
2006	37.1
2007	39.9
2008	43.6
2009	43.6
2010	65.5

2-10 海洋船舶工业增加值
Added Value of Marine Shipbuilding Industry

单位：亿元 (100 million yuan)

年 份 Year	增加值 Added Value
2001	109.3
2002	117.4
2003	152.8
2004	204.1
2005	275.5
2006	339.5
2007	524.9
2008	742.6
2009	986.5
2010	1 215.6

2-11 海洋化工业增加值
Added Value of Marine Chemical Industry

单位：亿元 (100 million yuan)

年 份 Year	增加值 Added Value
2001	64.7
2002	77.1
2003	96.3
2004	151.5
2005	153.3
2006	440.4
2007	506.6
2008	416.8
2009	465.3
2010	613.8

注：自2006年起部分地区统计产品品种增加。

Note: The data from 2006 include added kinds of statistical products in some regions.

2-12 海洋生物医药业增加值
Added Value of Marine Biomedicine Industry

单位：亿元 (100 million yuan)

年 份 Year	增加值 Added Value
2001	5.7
2002	13.2
2003	16.5
2004	19.0
2005	28.6
2006	34.8
2007	45.4
2008	56.6
2009	52.1
2010	83.8

2-13 海洋工程建筑业增加值
Added Value of Marine Engineering Architecture

单位：亿元 (100 million yuan)

年 份 Year	增加值 Added Value
2001	109.2
2002	145.4
2003	192.6
2004	231.8
2005	257.2
2006	423.7
2007	499.7
2008	347.8
2009	672.3
2010	874.2

2-14 海洋电力业增加值
Added Value of Marine Electric Power Industry

单位：亿元 (100 million yuan)

年 份 Year	增加值 Added Value
2001	1.8
2002	2.2
2003	2.8
2004	3.1
2005	3.5
2006	4.4
2007	5.1
2008	11.3
2009	20.8
2010	38.1

2-15 海水利用业增加值
Added Value of Seawater Utilization Industry

单位：亿元 (100 million yuan)

年 份 Year	增加值 Added Value
2001	1.1
2002	1.3
2003	1.7
2004	2.4
2005	3.0
2006	5.2
2007	6.2
2008	7.4
2009	7.8
2010	8.9

2-16　海洋交通运输业增加值
Added Value of Marine Communications and Transportation Industry

单位：亿元　　(100 million yuan)

年　份 Year	增加值 Added Value
2001	1 316.4
2002	1 507.4
2003	1 752.5
2004	2 030.7
2005	2 373.3
2006	2 531.4
2007	3 035.6
2008	3 499.3
2009	3 146.6
2010	3 785.8

2-17　滨海旅游业增加值
Added Value of Coastal Tourism

单位：亿元　　(100 million yuan)

年　份 Year	增加值 Added Value
2001	1 072.0
2002	1 523.7
2003	1 105.8
2004	1 522.0
2005	2 010.6
2006	2 619.6
2007	3 225.8
2008	3 766.4
2009	4 352.3
2010	5 303.1

2-18 沿海地区海洋生产总值
Gross Ocean Product by Coastal Regions

地区 Region	海洋生产总值（亿元） Gross Ocean Product (100 million yuan)				海洋生产总值占沿海地区生产总值比重（%） Proportion of the Gross Ocean Product in the Gross Regional Product（%）
		第一产业 Primary Industry	第二产业 Secondary Industry	第三产业 Tertiary Industry	
合计 Total	**39 572.7**	**2 008.0**	**18 935.0**	**18 629.8**	**16.1**
天津 Tianjin	3 021.5	6.1	1 979.7	1 035.7	32.8
河北 Hebei	1 152.9	47.1	653.8	452.1	5.7
辽宁 Liaoning	2 619.6	315.8	1 137.1	1 166.7	14.2
上海 Shanghai	5 224.5	3.7	2 059.6	3 161.1	30.4
江苏 Jiangsu	3 550.9	162.6	1 927.1	1 461.2	8.6
浙江 Zhejiang	3 883.5	286.7	1 763.3	1 833.6	14.0
福建 Fujian	3 682.9	317.7	1 602.5	1 762.7	25.0
山东 Shandong	7 074.5	444.0	3 552.2	3 078.3	18.1
广东 Guangdong	8 253.7	194.0	3 920.0	4 139.6	17.9
广西 Guangxi	548.7	100.4	223.1	225.2	5.7
海南 Hainan	560.0	129.9	116.6	313.5	27.1

2-19 沿海地区海洋生产总值构成
Composition of Gross Ocean Product by Coastal Regions

单位：% (%)

地 区 Region	海洋生产总值 Gross Ocean Product	第一产业 Primary Industry	第二产业 Secondary Industry	第三产业 Tertiary Industry
合 计 Total	**100.0**	**5.1**	**47.8**	**47.1**
天 津 Tianjin	100.0	0.2	65.5	34.3
河 北 Hebei	100.0	4.1	56.7	39.2
辽 宁 Liaoning	100.0	12.1	43.4	44.5
上 海 Shanghai	100.0	0.1	39.4	60.5
江 苏 Jiangsu	100.0	4.6	54.3	41.2
浙 江 Zhejiang	100.0	7.4	45.4	47.2
福 建 Fujian	100.0	8.6	43.5	47.9
山 东 Shandong	100.0	6.3	50.2	43.5
广 东 Guangdong	100.0	2.4	47.5	50.2
广 西 Guangxi	100.0	18.3	40.7	41.0
海 南 Hainan	100.0	23.2	20.8	56.0

2-20 沿海地区海洋及相关产业增加值
Added Values of Marine and Related Industries by Coastal Regions

单位：亿元 (100 million yuan)

地 区 Region	合 计 Total	海洋产业 Marine Industry	主要海洋产业 Major Marine Industry	海洋科研教育管理服务业 Industries of Marine Scientific Research, Education, Management and Service	海洋相关产业 Ocean-related Industries
合 计 Total	**39 572.7**	**22 831.0**	**16 187.8**	**6 643.1**	**16 741.7**
天 津 Tianjin	3 021.5	1 666.5	1 519.8	146.7	1 355.1
河 北 Hebei	1 152.9	593.8	525.5	68.3	559.2
辽 宁 Liaoning	2 619.6	1 640.6	1 288.7	351.9	979.0
上 海 Shanghai	5 224.5	3 035.1	1 989.6	1 045.6	2 189.3
江 苏 Jiangsu	3 550.9	1 953.9	1 481.6	472.2	1 597.1
浙 江 Zhejiang	3 883.5	2 183.1	1 550.3	632.8	1 700.4
福 建 Fujian	3 682.9	1 931.9	1 379.1	552.8	1 751.1
山 东 Shandong	7 074.5	4 024.5	2 960.4	1 064.2	3 050.0
广 东 Guangdong	8 253.7	5 066.0	2 935.1	2 130.8	3 187.7
广 西 Guangxi	548.7	338.8	277.7	61.1	210.0
海 南 Hainan	560.0	397.0	280.1	116.8	163.0

2-21 沿海地区海洋及相关产业增加值构成
Composition of the Added Value of Marine and Related Industries by Coastal Regions

单位：%　　(%)

地 区 Region	合 计 Total	海洋产业 Marine Industry			海洋相关产业 Ocean-related Industries
			主要海洋产业 Major Marine Industry	海洋科研教育管理服务业 Industries of Marine Scientific Research, Education, Management and Service	
合 计 **Total**	**100.0**	**57.7**	**40.9**	**16.8**	**42.3**
天 津 Tianjin	100.0	55.2	50.3	4.9	44.8
河 北 Hebei	100.0	51.5	45.6	5.9	48.5
辽 宁 Liaoning	100.0	62.6	49.2	13.4	37.4
上 海 Shanghai	100.0	58.1	38.1	20.0	41.9
江 苏 Jiangsu	100.0	55.0	41.7	13.3	45.0
浙 江 Zhejiang	100.0	56.2	39.9	16.3	43.8
福 建 Fujian	100.0	52.5	37.4	15.0	47.5
山 东 Shandong	100.0	56.9	41.8	15.0	43.1
广 东 Guangdong	100.0	61.4	35.6	25.8	38.6
广 西 Guangxi	100.0	61.7	50.6	11.1	38.3
海 南 Hainan	100.0	70.9	50.0	20.9	29.1

主要统计指标解释

1. 海洋经济 是开发、利用和保护海洋的各类产业活动以及与之相关联活动的总和。

2. 海洋生产总值 是海洋经济生产总值的简称，指按市场价格计算的沿海地区常住单位在一定时期内海洋经济活动的最终成果，是海洋产业和海洋相关产业增加值之和。

3. 海洋产业 是开发、利用和保护海洋所进行的生产和服务活动，包括海洋渔业、海洋油气业、海洋矿业、海洋盐业、海洋化工业、海洋生物医药业、海洋电力业、海水利用业、海洋船舶工业、海洋工程建筑业、海洋交通运输业、滨海旅游业等主要海洋产业以及海洋科研教育管理服务业。

4. 海洋科研教育管理服务业 是开发、利用和保护海洋过程中所进行的科研、教育、管理及服务等活动，包括海洋信息服务业、海洋环境监测预报服务、海洋保险与社会保障业、海洋科学研究、海洋技术服务业、海洋地质勘查业、海洋环境保护业、海洋教育、海洋管理、海洋社会团体与国际组织等。

5. 海洋相关产业 是指以各种投入产出为联系纽带，与主要海洋产业构成技术经济联系的上下游产业，涉及海洋农林业、海洋设备制造业、涉海产品及材料制造业、涉海建筑与安装业、海洋批发与零售业、涉海服务业等。

6. 海洋三次产业 我国的海洋三次产业划分如下：

海洋第一产业：是指海洋渔业中的海洋水产品、海洋渔业服务业，以及海洋相关产业中属于第一产业范畴的部门。

海洋第二产业：是指海洋渔业中海洋水产品加工、海洋油气业、海洋矿业、海洋盐业、海洋化工业、海洋生物医药业、海洋电力业、海水利用业、海洋船舶工业、海洋工程建筑业，以及海洋相关产业中属于第二产业范畴的部门。

海洋第三产业：是指除海洋第一、二产业以外的其他行业。第三产业包括：海洋交通运输业、滨海旅游业、海洋科研教育管理服务业，以及海洋相关产业中属于第三产业范畴的部门。

7. 海洋渔业 包括海水养殖、海洋捕捞、海洋渔业服务业和海洋水产品加工等活动。

8. 海洋油气业 是指在海洋中勘探、开采、输送、加工原油和天然气的生产活动。

9. 海洋矿业 包括海滨砂矿、海滨土砂石、海滨地热与煤矿及深海矿物等的采选活动。

10. 海洋盐业 是指利用海水生产以氯化钠为主要成分的盐产品的活动，包括采盐和盐加工。

11. 海洋船舶工业 是指以金属或非金属为主要材料，制造海洋船舶、海上固定及浮动装置的活动，以及对海洋船舶的修理及拆卸活动。

12. 海洋化工业 包括海盐化工、海水化工、海藻化工及海洋石油化工的化工产品生产活动。

13. 海洋生物医药业 是指以海洋生物为原料或提取有效成分，进行海洋药品与海洋保健品的生产加工及制造活动。

14. 海洋工程建筑业 是指在海上、海底和海岸所进行的用于海洋生产、交通、娱乐、防护等用途的建筑工程施工及其准备活动；包括海港建筑、滨海电站建筑、海岸堤坝建筑、海洋隧道桥梁建筑、海上油气田陆地终端及处理设施建造、海底线路管道和设备安装，不包括各部门、各地区的房屋建筑及房屋装修工程。

15. 海洋电力业 是指在沿海地区利用海洋能、海洋风能进行的电力生产活动。不包括沿海地区的火力发电和核力发电。

16. 海水利用业 是指对海水的直接利用和海水淡化活动，包括利用海水进行淡水生产和将海水应用于工业冷却用水和城市生活用水、消防用水等活动，不包括海水化学资源综合利用活动。

17. 海洋交通运输业 是指以船舶为主要工具从事海洋运输以及为海洋运输提供服务的活动，包括远洋旅客运输、沿海旅客运输、远洋货物运输、沿海货物运输、水上运输辅助活动、管道运输业、装卸搬运及其他运输服务活动。

18. 滨海旅游业 是指以海岸带、海岛及海洋各种自然景观、人文景观为依托的旅游经营、服务活动，主要包括：海洋观光游览、休闲娱乐、度假住宿、体育运动等活动。

Explanatory Notes on Main Statistical Indicators

1. Marine Economy is the summation of various types of industrial activities for developing, utilizing and protecting the ocean as well as the activities associated with there.

2. Gross Ocean Product is the short form of the gross output value of ocean economy, referring to the final result of marine economic activities of the permanent units in the coastal region within a given period calculated at the market price, and the sum total of the added values of the marine industries as the ocean-related industries.

3. Marine industry refers to the production as service activities for developing, utilizing and protecting the ocean, including major marine industries such as offshore oil and gas industry, marine mining industry, marine salt industry, marine chemical industry, marine biomedicine industry, marine electric power industry, seawater utilization industry, marine shipbuilding industry, marine engineering construction industry, marine communications and transportation industry, coastal tourism etc. as well as marine scientific research, education, management and service.

4. Marine Scientific Research, Education, Management and Service refer to the activities of scientific research, education, management and service carried out in the process of developing, utilizing and protecting the ocean, including marine information service industry, marine environment monitoring and forecasting service, marine insurance and social security industry, marine scientific research, marine technological service industry, ocean geological prospecting industry, marine environmental protection industry, marine education, marine management, marine social organization and international organizations etc.

5. Ocean-Related Industry refers to the lower and upper reaches enterprises that form a technical and economic link with the major marine industries, with various inputs and outputs as ties, involving marine agriculture and forestry, marine equipment manufacturing, ocean-related building and installation industry, marine wholesale and retail industry, ocean-related service industry etc.

6. Marine Three Industries Chinese marine three industries are divided as follows:

Marine primary industry: refers to the marine aquatic products, marine fishery service industry in the marine fishery as well as the sectors belonging to the primary industry category in the ocean-related industries.

Marine secondary industry: refers to the marine aquatic products processing industry in the marine fishery, offshore oil as gas industry, marine mining industry, marine salt industry, marine chemical industry, marine biomedicine industry, marine electric power industry, seawater utilization industry, marine shipbuilding industry, marine engineering construction industry, as well as the sectors belonging to the category of secondary industry in the ocean-related industries.

Marine tertiary Industry: refers to the industries other than the marine primary and secondary industries, including marine communications and transportation industry, coastal tourism, marine scientific research, education, management and service industry as well as the sectors belonging to the category of tertiary industry in the ocean-related industries.

7. Marine Fishery includes mariculture, marine fishing, marine fishery service industry and marine aquatic products processing, etc.

8. Offshore Oil and Gas Industry refers to the production activities of exploring, exploiting, transporting and processing crude oil and natural gas in the ocean.

9. Ocean Mining Industry includes the activities of extracting and dressing beach placers, beach soil and sand, submarine geothermal energy, and coal mining and deep-sea mining, etc.

10. Marine Salt Industry refers to the activity of producing the salt products with the sodium chloride as the main component by utilizing seawater, including salt extracting and processing.

11. Shipbuilding Industry refers to the activity of building ocean vessels, offshore fixed and floating equipment with metals or non-metals as main materials as well as repairing and dismantling ocean vessels.

12. Marine Chemical Industry includes the production activities of chemical products of sea salt, seawater, sea algal and marine petroleum chemical industries.

13. Marine Biomedicine Industry refers to the production, processing and manufacturing activities of marine medicines and marine health care products by using marine organisms as raw materials or extracting useful components therefrom.

14. Marine Engineering Building Industry refers to the architectural projects construction and its preparations in the sea, at the sea bottom and seacoast for such uses as marine production, transportation, recreation, protection, etc., including constructions of seaports, coastal power stations, coastal dykes, marine tunnels and bridges, land terminals of offshore oil and gas fields as well as building of processing facilities, and installation of submarine pipelines and equipment, but not the projects of house building and renovation.

15. Marine Electric Power Industry refers to the activities of generating electric power in the coastal region by making use of ocean energies and ocean wind energy. It does not include the thermal and nuclear power generation in the coastal area.

16. Seawater Utilization Industry refers to the activities of the direct use of sea water and the seawater desalination, including those of carrying out the production of desalination and applying the seawater as water for industrial cooling, urban domestic water, water for fire fighting etc., but not the activity of the multipurpose use of seawater chemical resources.

17. Marine Communications and Transportation Industry refers to the activities of carrying out and serving the sea transportations with vessels as main vehicles, including ocean-going passagers transportation, coastal passagers transportation, ocean-going cargo transportation, coastal cargo transportation, auxiliary activities of water transportation, pipeline transportation, loading, unloading and transport as well as other transportation service activities.

18. Coastal Tourism refers to the tourist business and service activities with the backing of coastal zone, sea islands as well as a variety of natural and human landscapes of the ocean, mainly including marine sightseeing, living a life of leisure and recreation, going on vocation and getting accommodation, sports, etc.

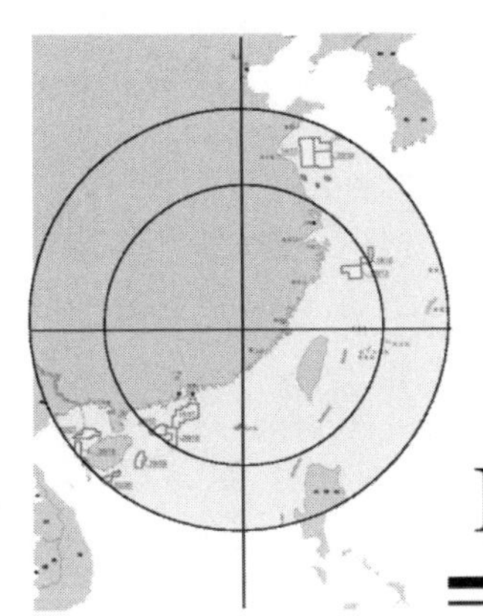

3 主要海洋产业活动

Major Marine Industrial Activities

3-1 全国海洋捕捞养殖产量
National Marine Catches and Mariculture Production

单位：吨 (t)

项　目　Item	2008	2009	2010
海水水产品产量 Total Seawater Aquatic Products	**27 844 671**	**28 805 399**	**27 975 312** ③
海洋捕捞产量 Marine Catches	**13 408 617** ①	**13 440 772** ①	**12 035 946** ②
按品种分　By Species			
鱼类　Fish	9 086 773	9 148 657	8 255 051
甲壳类　Crustacea	2 175 080	2 177 651	2 043 314
贝类　Shellfish	660 647	681 832	622 104
藻类　Algae	36 643	28 542	24 636
头足类　Cephalopoda	942 980	820 410	658 309
其他　Others	376 154	438 366	432 532
海水养殖产量 Mariculture Production	**14 436 054**	**15 364 627**	**14 823 008**
按品种分　By Species			
鱼类　Fish	884 310	938 601	808 171
甲壳类　Crustacea	1 003 893	1 102 028	1 061 096
贝类　Shellfish	10 742 153	11 398 288	11 082 321
藻类　Algae	1 483 395	1 549 394	1 541 322
其他　Others	322 303	376 316	330 098

注：①海洋捕捞产量分项中未包括辽宁远洋捕捞产量（2008，2009）。②海洋捕捞产量未包括远洋捕捞产量（2010）。③海水水产品产量为海洋捕捞产量、远洋捕捞产量及海水养殖产量之和。

Note: ① Data for Liaoning Deep-Sea fishing production is not included in the marine catches (2008, 2009).
② Data for Deep-Sea fishing production is not included in the marine catches.
③ The output of seawater aquatic products is the sum of the productions of marine fishing, deep-sea fishing and mariculture.

3-2 沿海地区海洋捕捞养殖产量
Marine Catches and Mariculture Production by Coastal Regions

单位：吨 (t)

地　区 Region	海洋捕捞产量 Marine Catches	远洋捕捞产量 Deep-Sea Fishing Production	海水养殖产量 Mariculture Production
合　计 Total	**12 035 946**	**887 979**	**14 823 008**
天　津 Tianjin	15 754	9 020	14 212
河　北 Hebei	253 292		329 308
辽　宁 Liaoning	1 007 398	175 274	2 314 694
上　海 Shanghai	21 531	99 933	
江　苏 Jiangsu	570 354	8 941	785 173
浙　江 Zhejiang	2 821 000	165 602	825 730
福　建 Fujian	1 908 468	180 524	3 038 990
山　东 Shandong	2 350 888	149 814	3 962 643
广　东 Guangdong	1 429 592	94 752	2 490 688
广　西 Guangxi	662 954	4 119	877 408
海　南 Hainan	994 715		184 162

3-3 沿海地区海洋原油产量
Output of Offshore Crude Oil by Coastal Regions

单位：万吨 (10 000 t)

地　区 Region	2008	2009	2010
合　计 Total	**3 421.13**	**3 698.19**	**4 709.98**
天　津 Tianjin	1 557.15	1 874.01	2 916.46
河　北 Hebei	189.62	200.30	221.19
辽　宁 Liaoning	22.80	15.00	13.01
上　海 Shanghai	15.34	9.68	8.80
山　东 Shandong	232.15	240.01	246.27
广　东 Guangdong	1 404.07	1 359.19	1 304.25

3-4 沿海地区海洋天然气产量
Output of Offshore Natural Gas by Coastal Regions

单位：万立方米 (10 000 m^3)

地　区 Region	2008	2009	2010
合　计 Total	**857 847**	**859 173**	**1 108 905**
天　津 Tianjin	140 111	143 002	186 089
河　北 Hebei	18 887	37 043	40 753
辽　宁 Liaoning	5 937	4 575	3 069
上　海 Shanghai	63 716	58 767	49 742
山　东 Shandong	16 798	16 157	12 874
广　东 Guangdong	612 398	599 629	816 378

3-5 海洋原油出口量及创汇额
Export Volume and Foreign-Exchange Earnings of Offshore Crude Oil by Coastal Regions

单位：万吨，万美元 (10 000 t, 10 000 US$)

地 区 Region	2008		2009		2010	
	出口量 Export Volume	创汇额 Foreign-Exchange Earnings	出口量 Export Volume	创汇额 Foreign-Exchange Earnings	出口量 Export Volume	创汇额 Foreign-Exchange Earnings
合 计 Total	**318.09**	**209 533**	**134.90**	**51 901**	**73.86**	**37 530**
天 津 Tianjin	186.02	113 629	66.15	24 679	48.28	23 640
广 东 Guangdong	132.07	95 904	68.75	27 222	25.58	13 890

3-6 海洋原油产量、出口量占全国原油产量、出口量比重
Proportion of Offshore Crude Oil Production and Export Volume in the National Total

年 份 Year	海洋原油产量 占全国原油产量比重（%） Proportion of Offshore Crude Oil Production in the National Total (%)	海洋原油出口量占 全国原油出口量比重（%） Proportion of Offshore Crude Oil Export Volume in the National Total (%)
2001	13.07	46.26
2002	14.40	54.95
2003	15.01	60.77
2004	16.16	83.37
2005	17.51	84.18
2006	17.54	89.53
2007	17.06	71.53
2008	17.96	76.46
2009	19.52	26.61
2010	23.20	24.38

3-7 沿海地区海洋矿业产量
Output of Marine Mining Industry by Coastal Regions

单位：吨 (t)

地 区 Region	产 量 Output		
	2008	2009	2010
合 计 Total	**48 085 689**	**55 906 609**	**34 225 357**
浙 江 Zhejiang	40 015 300	47 554 400	25 013 500
福 建 Fujian	2 063 800	2 082 500	2 287 100
山 东 Shandong	3 308 591	3 423 889	4 257 157
广 西 Guangxi	887 998	564 820	250 000
海 南 Hainan	1 810 000	2 281 000	2 417 600

3-8 沿海地区海盐产量
Productive Activities of Salt-Making Industry by Coastal Regions

单位：万吨 (10 000 t)

地　区 Region	海盐产量 Output of Sea Salt		
	2008	2009	2010
合　计 Total	**3 127.15**	**3 500.45**	**3 286.63**
天　津 Tianjin	235.96	227.56	204.40
河　北 Hebei	385.07	421.65	429.41
辽　宁 Liaoning	182.08	222.33	146.05
江　苏 Jiangsu	102.43	106.61	149.91
浙　江 Zhejiang	19.28	17.05	10.59
福　建 Fujian	40.13	53.14	29.99
山　东 Shandong	2 122.71	2 412.72	2 273.05
广　东 Guangdong	14.54	11.57	14.16
广　西 Guangxi	13.18	15.34	14.29
海　南 Hainan	11.77	12.48	14.78

3-9 沿海地区海洋化工产品产量
Output of Marine Chemical Products by Coastal Regions

单位：吨 (t)

地 区 Region	产品产量① Output		
	2008	2009	2010
合 计 Total	**14 558 214**	**12 915 773**	**11 499 913**
天 津 Tianjin	1 986 324	1 640 000	1 666 288
河 北 Hebei	771 750	820 340	73 960
辽 宁 Liaoning	771 622	446 095②	652 876
江 苏 Jiangsu	310 549	1 284 696	1 307 863
浙 江 Zhejiang	444 866	373 133②	494 421
福 建 Fujian	4 354 899	269 840	378 987
山 东 Shandong	5 197 364	7 240 769	6 349 018
广 东 Guangdong	720 840②	840 900②	576 500②

注：①数据为沿海地区部分海洋化工企业产品汇总数据;②为中国盐业总公司数据。

Note: ① The data collected from the products of part of the chemical enterprises in the coastal region.
② The data from the China Salt Industry Corporation.

3-10 沿海地区海洋生物医药产品产量*

Production of Marine Biomedicine Industry by Coastal Regions

产品名称 Name	计量单位 Unit	产品产量 Output
海参肽胶囊 Sea cucumber peptide nutrient capsule	万粒 10 000 pellets	150.00
藻日康 Zaorikang	盒 case	58 500.00
巨藻盖 Ju Zaogai	盒 case	96 700.00
螺旋藻粉 Spirulina Powder	吨 t	630.00
螺旋藻 Spirulina	万瓶 10 000 bottles	473.00
螺旋藻 Spirulina	万粒 10 000 pellets	2 281.00
螺旋藻胶囊 Spirulina Capsule	瓶 bottle	5 300.00
螺旋藻胶囊 Spirulina Capsule	吨 t	166.00
藻酸双酯钠片 Alginic Acid Diadipose Sodium Pill	万片 10 000 pills	10 237.00
鲨鱼肝油胶丸 Shark Liver Oil Pill	万粒 10 000 pellets	8 145.00
金枪鱼油胶丸 Tuna Oil Pill	万粒 10 000 pellets	2 856.00
卵磷脂胶丸 Lecithin Pill	万粒 10 000 pellets	2 323.00
鳕鱼肝油胶丸 Ling Liver Oil Pill	万粒 10 000 pellets	273.00
海藻植物胶囊 Algae Plant Capsule	万粒 10 000 pellets	225.00
蚝贝钙片 Oyster Shell Calcium Tablets	万片 10 000 pills	2 176.92
鱼油软胶囊 Fish Oil Soft Capsule	万粒 10 000 pellets	1 089.00
精炼鱼油 Refineed fish Oil	吨 t	1 215.00
鲨鱼硫酸软骨素 Shark Sulphate Chondroitin	吨 t	6.00
海蛇痹宁胶囊 Sea Snake Anti-rheumatic Capsule	万粒 10 000 pellets	304.20
维生素AD胶丸 Vitamin AD Capsule	万粒 10 000 pellets	59 244.00
维生素E胶丸 Vitamin E Capsule	万粒 10 000 pellets	38 400.00
鲎试剂 King Crab Tablet	万支 10 000 bottles	310.00
海珠喘息定片 Haizhu Methoxyphenamine Pill	万片 10 000 pills	15 948.00
珍珠粉末 Pearl Powder	万瓶 10 000 bottles	642.00
金牡感冒片 Jinmu Cold Cure Pill	万片 10 000 pills	435.05
碘[I-125]密封籽源 Iodine [I-125] sealed	粒 pellet	86 000.00
鳃腺炎减毒活疫苗 Parotitis toxicity reducing living vaccine	人/份 Person/portion	4 298 420.00

3-10 续表 continued

产品名称 Name	计量单位 Unit	产品产量 Output
甲壳素 Crustaceoxin	吨 t	1 834.00
海墨止血片 Haimo Haemostatic Pill	万盒 10 000 cases	9.00
海参保健系列产品 Sea cucumber health care series products	盒 case	5 000.00
海威口服液 Haiwei oral liquid	吨 t	428.00
依可新 Vitamin A and D Drops	吨 t	164.00
苯海因 Benzene Glycolylurea	吨 t	1 360.00
乙酰胺基噻二唑 Acetazolamide	吨 t	38.00
羟混苯 Hydroxy Mixed Benzene	吨 t	3 100.00
角鲨烯胶囊 Houndfish Alkene Capsule	万粒 10 000 pellets	110.00
D酯 D ester	吨 t	1 500.00
麝珠明目滴眼液 Shezhu eye-brightening drops	万瓶 10 000 bottles	142.00
碘酊 Iodo-tincture	万瓶 10 000 bottles	12.12
碘甘油 Iodoglycerol	万瓶 10 000 bottles	4.40
鱼石脂软膏 Ichthyol cream	万支 10 000 bottles	112.12
八宝惊风散 Babao infantile convulsions powder	万盒 10 000 cases	106.05
海藻酸钠 Sodium algrnic acid	吨 t	132 323.00
碘 Iodine	吨 t	1 510.00
盐酸氨基葡萄糖 Aminoglucose hydrochloride	吨 t	2 000.00
海狗油软胶囊 Fur seal oil soft capsule	瓶 bottles	39 516.00
金枪鱼软胶囊 Tuna soft capsule	万粒 10 000 pellets	193.00
甘露醇 Mannitol	万瓶 10 000 bottles	198.40
二硫基1，3，4噻二唑 Disulfenyl 1, 3, 4 thiadiazole	吨 t	292.00
肤疹宁软膏 Rash cream	吨 t	1.00
平安清消毒剂 Pinganqing disinfectant	吨 t	40.00
氨基葡萄糖盐酸盐 Aminoglucose hydrochloride	吨 t	679.00
保健品鱼油 Health care fish oil	吨 t	2 859.00
浓缩鱼油 Concentrated fish oil	吨 t	793.00
南海岸鳗钙系列产品 South coast eel-calcium	万盒 10 000 cases	220.00
伤科接骨片 Traumatologic osteopathic tablet	吨 t	166.00
龟鳖丸 Testudinate pill	吨 t	21.00

注：数据为沿海地区部分海洋生物医药产品汇总数据。

Note: The data collected from the products of part of the marine biomedicine enterprises in the coastal region.

3-11 沿海地区海洋修造船完工量
Production of the Marine Shipbuilding Industry by Coastal Regions

部门和地区 Sector and Region	修船完工量（艘） Ships Repaired (number)	造船完工量 Ships Built	
		艘 number	万综合吨 Comprehensive Tonnages (10 000 t)
总 计 Total	**10 066**	**2 387**	**6 397.41**
其 中: Including:			
中船工业集团公司 CSSC	561	171	1 544.09
中船重工集团公司 CSIC	628	85	921.00
按地区分: By Regions:			
天 津 Tianjin	187	20	20.77
河 北 Hebei	321	6	19.00
辽 宁 Liaoning	234	99	942.00
上 海 Shanghai	1 308	134	1 212.60
江 苏 Jiangsu	564	624	2 300.30
浙 江 Zhejiang	4 393	837	1 029.19
福 建 Fujian	1 715	342	133.56
山 东 Shandong	790	75	404.19
广 东 Guangdong	444	61	332.00
广 西 Guangxi	72	123	3.13
海 南 Hainan	38	65	0.67

3-12 沿海地区海洋货物运输量和周转量
Maritime Volume of Goods Transported and Turnover by Coastal Regions

单位：万吨，亿吨千米 (10 000 t, 100 million t-km)

地 区 Region	货运量 Volume of Goods Transported	沿 海 Coastal	远 洋 Oceangoing	货物周转量 Volume of Goods Turnover	沿 海 Coastal	远 洋 Oceangoing
全国总计 National Total	**190 370**	**132 316**	**58 054**	**62 892**	**16 893**	**45 999**
天 津 Tianjin	11 912	4 169	7 743	9 324	768	8 556
河 北 Hebei	2 150	2 150		441	441	
辽 宁 Liaoning	10 423	4 986	5 437	5 696	746	4 950
上 海 Shanghai	43 628	28 456	15 172	18 588	4 053	14 535
江 苏 Jiangsu	12 988	8 911	4 077	3 402	885	2 517
浙 江 Zhejiang	36 523	34 941	1 582	5 098	4 059	1 039
福 建 Fujian	14 032	12 526	1 506	2 207	1 757	450
山 东 Shandong	11 459	5 802	5 657	3 896	545	3 351
广 东 Guangdong	20 414	11 983	8 431	3 360	1 669	1 691
广 西 Guangxi	2 952	2 605	347	479	464	15
海 南 Hainan	7 612	7 329	283	848	801	47
其 他 Others	16 277	8 458	7 819	9 553	705	8 848

3-13 沿海地区海洋旅客运输量和周转量
Maritime Volume of Passenger Traffic and Turnover by Coastal Regions

单位：万人，亿人千米 (10 000 persons, 100 million person-km)

地 区 Region	客运量 Passenger Traffic	沿 海 Coastal	远 洋 Oceangoing	旅客周转量 Passenger Turnover Volume	沿 海 Coastal	远 洋 Oceangoing
全国总计 National Total	**10 218**	**9 355**	**863**	**42.73**	**32.74**	**9.99**
天 津 Tianjin	1		1	0.18		0.18
辽 宁 Liaoning	490	479	11	6.39	5.85	0.54
上 海 Shanghai	504	504		5.42	5.42	
江 苏 Jiangsu	19	19		0.82	0.82	
浙 江 Zhejiang	2 472	2 472		5.27	5.27	
福 建 Fujian	1 215	1 135	80	1.77	1.33	0.44
山 东 Shandong	2 321	2 234	87	11.63	7.98	3.65
广 东 Guangdong	1 956	1 274	682	7.91	2.79	5.12
广 西 Guangxi	132	130	2	0.70	0.65	0.05
海 南 Hainan	1 108	1 108		2.65	2.65	

3-14 沿海港口客货吞吐量
Passengers Leaving and Arriving and Cargo Handled at Coastal Seaports

单位：万吨，万人次 (10 000 t, 10 000 person-times)

地 区 Region	货物吞吐量 Cargo Handled	#外 贸 Foreign Trade	旅客吞吐量 Passenger Leaving & Arriving	#离 港 Leaving
合 计 Total	**564 464**	**228 810**	**7 332**	**3 694**
天 津 Tianjin	41 325	20 709	23	12
河 北 Hebei	60 344	13 607	6	3
辽 宁 Liaoning	67 790	16 845	631	310
上 海 Shanghai	56 320	30 225	168	85
江 苏 Jiangsu	13 847	7 871	14	7
浙 江 Zhejiang	78 846	29 375	1 065	530
福 建 Fujian	32 687	12 789	995	499
山 东 Shandong	86 421	49 073	1 145	568
广 东 Guangdong	105 299	39 231	2 109	1 091
广 西 Guangxi	11 923	7 195	27	13
海 南 Hainan	9 662	1 891	1 149	577

3-15 沿海国际标准集装箱运量
Volume of International Standardized Containers Traffic at Coastal Seaports

单位：万标准箱，万吨 (10 000 TEU, 10 000 t)

地 区 Region	2008		2009		2010	
	箱 数 Containers	重 量 Weight	箱 数 Containers	重 量 Weight	箱 数 Containers	重 量 Weight
合 计 Total	**3 219**	**34 975**	**3 011**	**34 578**	**3 794**	**42 267**
天 津 Tianjin	12	112	12	68	12	177
河 北 Hebei	2	15	1	20	1	24
辽 宁 Liaoning	16	146	39	393	44	444
上 海 Shanghai	1 511	18 418	1 332	17 710	1 623	20 561
江 苏 Jiangsu	190	1 670	331	2 702	392	3 283
浙 江 Zhejiang	53	669	80	1 529	102	1 515
福 建 Fujian	145	1 745	205	3 272	231	3 902
山 东 Shandong	170	1 746	132	1 416	166	2 227
广 东 Guangdong	795	6 772	704	5 063	913	6 812
广 西 Guangxi	145	1 209	49	675	77	992
海 南 Hainan	20	285	22	295	100	317
其 他 Others	160	2 188	106	1 435	134	2 011

注：国际标准集装箱运量数据包含内河集装箱运量。

Note: The data in this table include the volume of containers transported in the inland rivers.

3-16 沿海港口国际标准集装箱吞吐量
International Standardized Containers Handled at Coastal Seaports

单位：万标准箱，万吨 (10 000TEU, 10 000 t)

地区 Region	2008		2009		2010	
	箱数 Containers	重量 Weight	箱数 Containers	重量 Weight	箱数 Containers	重量 Weight
合计 Total	**11 673**	**114 692**	**11 020**	**114 415**	**13 145**	**137 076**
天津 Tianjin	850	8 591	870	8 871	1 009	10 916
河北 Hebei	65	888	57	869	62	997
辽宁 Liaoning	744	10 372	812	11 718	969	15 666
上海 Shanghai	2 801	25 992	2 500	24 619	2 907	27 992
江苏 Jiangsu	301	2 816	305	2 872	391	3 810
浙江 Zhejiang	1 148	9 357	1 118	9 839	1 404	12 276
福建 Fujian	743	8 709	716	9 001	867	10 743
山东 Shandong	1 321	14 052	1 312	13 817	1 531	15 189
广东 Guangdong	3 620	32 779	3 236	31 432	3 868	37 413
广西 Guangxi	34	523	35	555	56	926
海南 Hainan	46	613	59	822	82	1 147

3-17 沿海城市国内旅游人数
Domestic Visitors by Coastal Cities

单位：万人次 (10 000 person-times)

城 市 City		2007	2008	2009
合 计	**Total**	**65 875**	**74 699**	**91 807**
天 津	**Tianjin**	6 018	7 004	5 537
河 北	**Hebei**	2 613	2 553	3 362
唐 山	Tangshan	762	957	1 226
秦皇岛	Qinhuangdao	1 510	1 227	1 638
沧 州	Cangzhou	341	369	498
辽 宁	**Liaoning**	6 150	7 735	9 800
大 连	Dalian	2 480	3 000	3 412
丹 东	Dandong	1 100	1 420	1 897
锦 州	Jinzhou	680	870	1 191
营 口	Yingkou	450	585	820
盘 锦	Panjin	760	980	1 270
葫芦岛	Huludao	680	880	1 211
上 海	**Shanghai**	10 210	11 006	12 361
江 苏	**Jiangsu**	2 686	3 146	3 655
南 通	Nantong	1 072	1 275	1 483
连云港	Lianyungang	911	1 065	1 210
盐 城	Yancheng	703	806	962
浙 江	**Zhejiang**	16 879	19 229	21 948
杭 州	Hangzhou	4 112	4 552	5 094
宁 波	Ningbo	3 074	3 465	3 962
温 州	Wenzhou	2 192	2 547	2 931
嘉 兴	Jiaxing	1 850	2 140	2 492
绍 兴	Shaoxing	2 192	2 435	2 851
舟 山	Zhoushan	1 285	1 495	1 731
台 州	Taizhou	2 174	2 595	2 887
福 建	**Fujian**			6 938
福 州	Fuzhou			1 938
厦 门	Xiamen			1 788
莆 田	Putian			632
泉 州	Quanzhou			1 171
漳 州	Zhangzhou			835
宁 德	Ningde			574

3-17 续表 continued

城 市 City		2007	2008	2009
山 东	**Shandong**	9 927	11 490	13 580
青 岛	Qingdao	3 259	3 390	3 903
东 营	Dongying	329	428	515
烟 台	Yantai	1 999	2 346	2 763
潍 坊	Weifang	1 414	1 869	2 313
威 海	Weihai	1 358	1 586	1 839
日 照	Rizhao	1 226	1 451	1 724
滨 州	Binzhou	342	421	523
广 东	**Guangdong**	9 238	10 170	11 696
广 州	Guangzhou	2 727	2 916	3 286
深 圳	Shenzhen	1 729	1 790	1 944
珠 海	Zhuhai	553	829	911
汕 头	Shantou	542	604	668
江 门	Jiangmen	619	728	746
湛 江	Zhanjiang	142	165	462
茂 名	Maoming	179	171	201
惠 州	Huizhou	586	675	799
汕 尾	Shanwei	189	219	239
阳 江	Yangjiang	238	240	273
东 莞	Dongguan	960	994	1 191
中 山	Zhongshan	442	463	504
潮 州	Chaozhou	193	219	274
揭 阳	Jieyang	139	157	198
广 西	**Guangxi**	1 097	1 191	1 630
北 海	Beihai	601	695	810
防城港	Fangchenggang	194	150	418
钦 州	Qinzhou	302	346	402
海 南	**Hainan**	1 057	1 175	1 299
海 口	Haikou	571	622	662
三 亚	Sanya	486	553	637

注：数据来源于《中国区域经济统计年鉴》（2010）。

Note：The data come from *China Statistical Yearbook For Regional Economy (2010)*.

3-18 主要沿海城市国际旅游（外汇）收入
International Tourism (Foreign Exchange) Receipts in Major Coastal Cities

单位：万美元 (10 000 US$)

城 市	City	2008	2009	2010
天 津	Tianjin	100 139	118 264	141 951
秦皇岛	Qinhuangdao	10 169	11 909	12 022
大 连	Dalian	65 835	72 748	80 386
上 海	Shanghai	497 172	474 402	634 092
南 通	Nantong	28 368	30 933	36 066
连云港	Lianyungang	7 959	9 173	10 747
杭 州	Hangzhou	129 610	137 995	169 008
宁 波	Ningbo	46 874	48 650	59 066
温 州	Wenzhou	16 109	17 797	21 115
福 州	Fuzhou	65 750	77 400	84 300
厦 门	Xiamen	81 865	90 194	108 552
泉 州	Quanzhou	65 980	64 771	66 737
漳 州	Zhangzhou	11 653	12 685	15 455
青 岛	Qingdao	50 030	55 178	60 103
烟 台	Yantai	26 708	31 081	37 707
威 海	Weihai	13 734	16 083	19 151
广 州	Guangzhou	313 035	362 396	466 127
深 圳	Shenzhen	270 399	276 026	315 896
珠 海	Zhuhai	94 823	102 670	122 339
汕 头	Shantou	6 491	4 910	5 016
湛 江	Zhanjiang	1 891	2 237	2 716
中 山	Zhongshan	22 702	20 434	27 591
北 海	Beihai	1 559	1 721	2 173
海 口	Haikou	3 702	3 089	3 756
三 亚	Sanya	26 255	19 497	24 504

3-19 主要沿海城市接待入境旅游者人数
Number of Inbound Tourist Arrivals in Major Coastal Cities

单位：人次 (person-time)

城 市 City		2008	2009	2010
天 津	Tianjin	1 220 392	1 410 244	1 660 682
秦皇岛	Qinhuangdao	187 267	224 206	242 337
大 连	Dalian	950 045	1 050 043	1 166 020
上 海	Shanghai	5 264 727	5 333 935	7 337 216
南 通	Nantong	280 037	299 866	355 133
连云港	Lianyungang	90 922	100 076	116 663
杭 州	Hangzhou	2 213 319	2 304 045	2 757 147
宁 波	Ningbo	756 776	800 548	951 680
温 州	Wenzhou	318 230	329 734	391 587
福 州	Fuzhou	639 375	629 314	698 607
厦 门	Xiamen	1 045 309	1 281 907	1 551 865
泉 州	Quanzhou	630 776	626 721	770 457
漳 州	Zhangzhou	192 709	219 382	247 469
青 岛	Qingdao	800 836	1 000 670	1 080 511
烟 台	Yantai	352 090	400 901	472 023
威 海	Weihai	288 277	322 676	372 646
广 州	Guangzhou	6 124 801	6 894 044	8 147 900
深 圳	Shenzhen	8 695 727	8 963 697	10 206 000
珠 海	Zhuhai	2 862 150	2 978 421	3 251 400
汕 头	Shantou	139 313	123 161	133 800
湛 江	Zhanjiang	32 994	50 665	103 200
中 山	Zhongshan	651 753	475 769	480 600
北 海	Beihai	55 702	61 437	73 008
海 口	Haikou	136 128	103 206	132 877
三 亚	Sanya	511 476	317 833	415 939

3-20 主要沿海城市接待入境旅游者情况
Breakdown of Inbound Tourists in Major Coastal Cities

单位：人次，人天 (person-time, night)

城市 City	外国人 Foreigners		香港同胞 Hong Kong	
	人次数 Arrivals	人天数 Nights	人次数 Arrivals	人天数 Nights
天 津 Tianjin	1 530 461	5 674 002	48 471	557 857
秦皇岛 Qinhuangdao	228 976	768 351	5 825	13 385
大 连 Dalian	1 047 343	3 494 095	54 232	178 282
上 海 Shanghai	5 931 211	20 751 747	623 969	2 014 754
南 通 Nantong	315 992	1 711 811	14 894	88 634
连云港 Lianyungang	93 119	572 617	7 008	55 902
杭 州 Hangzhou	1 878 528	5 297 569	332 626	888 155
宁 波 Ningbo	538 932	1 743 007	158 299	404 202
温 州 Wenzhou	309 245	794 485	32 477	71 535
福 州 Fuzhou	389 198	3 211 353	89 803	639 364
厦 门 Xiamen	511 096	2 468 227	155 878	651 473
泉 州 Quanzhou	90 171	534 579	470 663	2 399 116
漳 州 Zhangzhou	55 864	224 470	81 713	294 288
青 岛 Qingdao	826 628	2 375 719	113 374	283 614
烟 台 Yantai	390 028	1 696 008	22 400	58 906
威 海 Weihai	355 989	875 630	2 655	5 773
广 州 Guangzhou	2 944 400	7 787 900	4 183 400	8 467 200
深 圳 Shenzhen	1 675 800	3 634 600	8 017 300	16 244 800
珠 海 Zhuhai	567 300	1 691 700	1 074 800	2 447 800
汕 头 Shantou	82 000	188 600	42 300	80 400
湛 江 Zhanjiang	35 000	94 800	54 000	139 800
中 山 Zhongshan	134 800	681 100	243 500	841 800
北 海 Beihai	37 249	58 578	21 904	33 088
海 口 Haikou	74 461	134 847	23 189	37 219
三 亚 Sanya	324 486	1 173 548	69 122	138 185

3-20 续表 continued

城 市 City	澳门同胞 Macao		台湾同胞 Taiwan Province	
	人次数 Arrivals	人天数 Nights	人次数 Arrivals	人天数 Nights
天　津 Tianjin	8 108	35 559	73 642	537 710
秦皇岛 Qinhuangdao	576	1 345	6 960	15 573
大　连 Dalian	1 579	5 572	62 866	204 972
上　海 Shanghai	40 043	140 003	741 993	2 871 741
南　通 Nantong	1 647	7 757	22 600	142 204
连云港 Lianyungang	3 029	7 328	13 507	99 787
杭　州 Hangzhou	22 095	49 434	523 898	1 864 623
宁　波 Ningbo	52 427	142 863	202 022	491 909
温　州 Wenzhou	11 402	16 521	38 463	88 328
福　州 Fuzhou	4 759	24 579	214 847	525 238
厦　门 Xiamen	5 574	24 589	879 317	2 506 787
泉　州 Quanzhou	38 605	172 069	171 018	633 017
漳　州 Zhangzhou	8 008	22 662	101 884	297 127
青　岛 Qingdao	28 133	86 056	112 376	287 356
烟　台 Yantai	12 513	24 309	47 082	152 861
威　海 Weihai	677	1 505	13 325	37 481
广　州 Guangzhou	456 800	933 100	563 300	1 391 800
深　圳 Shenzhen	48 400	84 200	464 500	1 113 200
珠　海 Zhuhai	742 200	1 777 300	867 100	2 190 700
汕　头 Shantou	500	800	9 000	19 900
湛　江 Zhanjiang	4 800	8 700	9 400	20 900
中　山 Zhongshan	63 900	209 900	38 400	150 600
北　海 Beihai	3 356	5 194	10 499	16 122
海　口 Haikou	911	1 557	34 316	52 205
三　亚 Sanya	3 013	6 592	19 318	37 026

主要统计指标解释

1. 海洋捕捞产量 凡是从海洋里捕捞的天然生长的水产品产量为捕捞产量。

2. 海水养殖产量 凡是从人工投放苗种或天然纳苗并进行人工饲养管理的海水养殖水域中捕捞的水产品产量为海水养殖产量。

3. 远洋捕捞产量 由各远洋渔业企业和各生产单位按我国远洋渔业项目管理办法组织的远洋渔船（队）在非我国管辖海域（外国专属经济区水域或公海）捕捞的水产品产量。中外合资、合作渔船捕捞的水产品只统计按协议应属于中方所有的部分。

4. 原油产量 是按净原油量来计算的，能直接用于销售和生产自用的原油量。目前海洋石油系统原油产量计算方法采用倒算法。

原油产量=销售量+期末库存量-期初库存量+海上平台及陆地终端处理厂自用量。

5. 天然气产量 指进入集输管网的销售量和就地利用的全部气量。

天然气产量=外输（销）量+企业自用量

6. 海洋原油出口量 指销往国外的产品数量。

7. 海洋原油出口创汇额 指产品销往国外的归中方所有的全部外汇收入。以美元或万美元表示。

8. 造船综合吨 等于以计量单位载重吨和满载排水量吨的民用船舶的吨位数之和。

9. 货运量 指经船舶实际运送的货物重量，按到达量统计。

10. 货物周转量 指实际运送的货物与其运送距离的乘积。

11. 集装箱运量 既包含货重，也包含箱重。箱重系指承运租用的空箱重量凡有运费收入的空箱，其重量应统计为运量，按空箱 1 吨为货运量 1 吨计算；若无收入，所承运的空箱一律不作运量统计。

12. 旅客周转量 指实际运送的旅客人数与其运送距离的乘积。

13. 国际旅游外汇收入 入境游客在中国（大陆）境内旅行、游览过程中用于交通、参观游览、住宿、餐饮、购物、娱乐等的全部花费。

14. 接待人次数 指报告期内我国接待游客人数。游客按出游地分为入境游客和国内游客，按出游时间分为旅游者（过夜游客）和一日游游客（不过夜游客）。

15. 接待人天数 指过夜旅游者的停留天数。

16. 外国人 指外国国籍的人，加入外籍的中国血统华人也计入外国人。

17. 港澳台同胞 指居住在我国香港特别行政区、澳门特别行政区和台湾省的中国同胞。

Explanatory Notes on Main Statistical Indicators

1. Marine Catches refers to the output of the naturally growing aquatic products caught from the sea.

2. Mariculture Production refers to the output of aquatic products whose young are artificially released or naturally collected, and raised and managed artificially, and which are caught from the waters of mariculture.

3. Deep-Sea Fishing Production refers to the output of aquatic products caught in the non-Chinese jurisdictional sea areas (foreign EEE or high sea) by the distant fishing vessels (fleet) organized by various distant fishing businesses and production units according to the management measures of the China distant fishing projects. The aquatic products caught by the Chinese-foreign joint ventures' and cooperative fishing vessels are counted only for the part owned by the Chinese side according to the agreement.

4. Output of Crude Oil is calculated on the basis of the net amount of crude oil, i.e., the amount of crude oil that may be directly used for sale and for the production itself.

Output of crude oil = Volume of sales + Reserves at the end of the period - Reserves at the beginning of the period +Amount for self-use on the platforms and in the terminal processing plants on land.

5. Output of Natural Gas refers to the total gas volume of the sales volume entering the oil collecting and transport pipeline network and that used locally.

Output of natural gas = Volume of sales or transport to other areas + Volume used by the enterprise itself

6. Export Volume of Offshore Crude Oil refers to the amount of products for sale abroad.

7. Foreign Exchange Earnings of Offshore Crude Oil refer to the total foreign exchange income from oil (gas) products for sale abroad which is owned by the Chinese side.

8. Comprehensive Tonnages of Shipbuilding refers to the sum of tonnage of civilian vessels with the deadweight capacity and full-load displacement as measured.

9. Freight Traffic refers to the weight of cargoes actually transported by vessels, which is counted according to the volume of arrival.

10. Cargoes Turnover Volume refers to the product of the actually transported cargoes and the transport distance.

11. Freight Volume of Containers includes the weight of both cargoes and container boxes. The weight of container boxes refers to the weight of empty containers rented for transport or having freight income and should be included in the freight volume, one ton of empty boxes equalling to one ton of freight volume. The empty boxes which have no income for transportation are not included in the freight volume.

12. Passenger Turnover Volume refers to the product of the number of passengers actually transported and the shipping distance.

13. International Tourism (Foreign Exchange) Receipts refer to the total expenditure made by inbound visitors within the territory of China (the mainland) in their course of travel on transport, tours and sightseeing, lodging, food and beverage, shopping, entertainment, etc.

14. Number of Person-Times Received refers to the number of visitors received by China in the period reported. Visitors are divided into inbound visitors and domestic visitors by origin of the travel, and tourists (overnight visitors) and same-day visitors by their length of stay.

15. Number of the Days of Stay refers to the number of the days of stay of tourists.

16. Foreigners refer to the people with foreign nationality, including foreign nationals of Chinese descent.

17. Compatriots from Hong Kong, Macao and Taiwan Province refer to the Chinese compatriots living in the Hong Kong Special Administrative Region, the Macao Special Administrative Region and Taiwan Province.

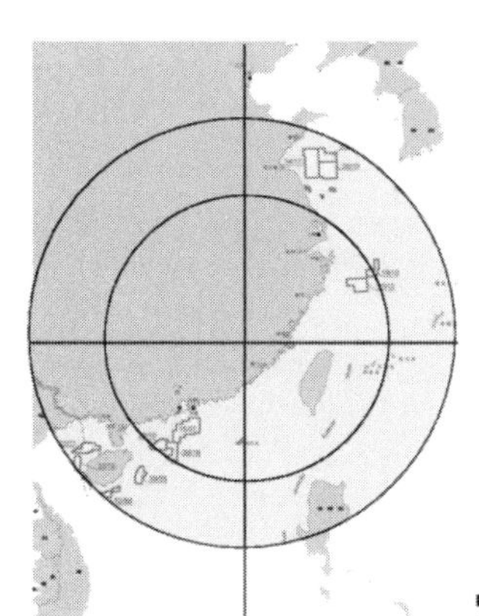

4

主要海洋产业生产能力

Production Capacity of Major Marine Industries

4-1 浅海滩涂海湾可养殖面积
Cultivable Area of Shellow Sea, Tidal Flat and Bay

单位：千公顷 (1 000 hm^2)

地 区	Region	海水可养殖面积 Cultivatable Marine Area	浅 海 Shallow Sea	滩 涂 Tidal Flat	港 湾 Bay
合 计	**Total**	**2 599.67**	**1 622.56**	**796.56**	**180.55**
天 津	Tianjin	18.49	10.00	8.49	
河 北	Hebei	111.37	49.66	61.70	
辽 宁	Liaoning	725.84	590.44	92.45	42.95
上 海	Shanghai	3.22		3.22	
江 苏	Jiangsu	139.00	7.87	130.96	0.17
浙 江	Zhejiang	101.46	36.30	57.39	7.77
福 建	Fujian	184.94	77.39	100.76	6.79
山 东	Shandong	358.21	131.68	173.41	53.12
广 东	Guangdong	835.67	664.00	120.00	51.67
广 西	Guangxi	31.95	6.78	22.09	3.08
海 南	Hainan	89.52	48.43	26.09	15.00

4-2　沿海地区海水养殖面积
Mariculture Area by Coastal Regions

单位：公顷　　(hm²)

地　区 Region	海水养殖面积 Mariculture Area
合　计 **Total**	**2 080 880**
天　津 Tianjin	3 982
河　北 Hebei	123 810
辽　宁 Liaoning	763 101
上　海 Shanghai	
江　苏 Jiangsu	192 426
浙　江 Zhejiang	93 905
福　建 Fujian	137 636
山　东 Shandong	500 946
广　东 Guangdong	199 258
广　西 Guangxi	51 287
海　南 Hainan	14 529

4-3 海洋油田生产井情况
Survey of Offshore Oilfield Production Wells

单位：口 (number)

地区 Region	合计 Total	采油井 Oil Wells	采气井 Gas Wells	注水井 Injection Wells	其他井 Others
合计 Total	**4 473**	**3 405**	**181**	**830**	**57**
天津 Tianjin	2 772	2 083	103	541	45
河北 Hebei	544	422		122	
辽宁 Liaoning	248	191	10	35	12
上海 Shanghai	32	12	20		
山东 Shandong	516	376	8	132	
广东 Guangdong	361	321	40		

4-4 海洋石油勘探情况
Work Volume of Offshore Oil Exploration

地区 Region	地震测线 Seismic Line		钻井（口） Drilling (Well)	
	二维（千米） Two Dimensions (km)	三维（平方千米） Three Dimensions (km^2)	预探井 Wildcat Wells	评价井 Appraisal Wells
合计 Total	**23 381**	**13 649**	**110**	**66**
天津 Tianjin		6 185	35	22
河北 Hebei	378		23	10
辽宁 Liaoning			2	2
上海 Shanghai	2 578	554	7	2
其中：合作 Including: Cooperative		554	7	2
山东 Shandong	405		8	10
广东 Guangdong	20 020	6 910	35	20
其中：合作 Including: Cooperative		2 208	9	2

4-5 沿海地区盐田面积和海盐生产能力
Salt Pan Area and Sea Salt Production Capacity by Coastal Regions

地 区 Region	盐田总面积（公顷）Total Area of Salt Pan (hm^2)		生产面积（公顷）Production Area (hm^2)		年末海盐生产能力（万吨）Year-End Capacity of Sea Salt Production (10 000 t)	
	2009	2010	2009	2010	2009	2010
合 计 Total	**465 974**	**473 068**	**324 453**	**332 099**	**4 194.04**	**3 960.00**
天 津 Tianjin	32 501	30 512	30 866	29 050	196.39	180.00
河 北 Hebei	78 613	80 359	67 871	65 243	527.46	476.00
辽 宁 Liaoning	45 146	39 790	39 340	33 533	300.48	234.00
江 苏 Jiangsu	70 218	65 634	24 729	19 204	126.00	160.00
浙 江 Zhejiang	3 643	3 355	2 988	2 703	16.96	16.00
福 建 Fujian	6 029	5 607	5 501	4 893	50.00	45.00
山 东 Shandong	212 006	230 122	142 170	166 962	2 927.55	2 800.00
广 东 Guangdong	10 579	10 052	6 368	6 014	16.11	19.00
广 西 Guangxi	3 552	3 949	1 778	1 655	14.59	11.00
海 南 Hainan	3 687	3 688	2 842	2 842	18.50	19.00

4-6 沿海地区风能发电能力（2009年）
Wind Power Production by Coastal Regions, 2009

单位：万千瓦 (10 000 kW)

地　区 Region	风能年发电能力 Annual Wind Power Generation Capacity
合　计 Total	**470.72**
辽　宁 Liaoning	50.27
河　北 Hebei	19.80
上　海 Shanghai	14.19
江　苏 Jiangsu	107.68
浙　江 Zhejiang	23.42
福　建 Fujian	56.73
山　东 Shandong	121.83
广　东 Guangdong	56.93
广　西 Guangxi	0.25
海　南 Hainan	19.62

4-7　主要潮汐电站分布情况
Distribution of Major Tidal Power Stations

电站名称 Name	运行情况 Status of operation	装机容量（千瓦） Capacity (kW)
江厦潮汐试验电站	1970年开始建造，1980年投入使用，运行至今。	3 900
Jiangxia Experimental Tidal Power Station	It began construction in 1970, was put into use in 1980, and has been in operation so far.	
海山潮汐电站	1972年开始建造，1975年投入使用，运行至今。	250
Haishan Tidal Power Station	It began construction in 1972, was put into use in 1975, and has been in operation so far.	
岳浦潮汐电站	1970年开始建造，1978年停止运行。	
Yuepu Tidal Power Station	It began construction in 1970, stopped power generation in 1978.	
白沙口潮汐电站	1970年开始建造，1978年投入使用，2010年停止运行。	
Baishakou Tidal Power Station	It began construction in 1970, was put into use in 1978, stopped power generation in 2010.	
果子山潮汐电站	已经投入前期筹备工作阶段，未确定开工时间。	
Guozishan Tidal Power Station	It has been put into the first stage of preparation, and the openning date has not been fixed.	

4-8 主要海上活动船舶
Major Vessels Operating on the Sea

类 别 Type	艘数 （艘） Number of Vessels (unit)	总吨 （万吨位） grt (10 000 t)	净载重量 （万吨） Net Weight Tonnage (10 000 t)	载客量 （客位） Passenger Spaces (seat)	总功率 （千瓦） Total Power (kW)
一 、海洋生产用船 **Vessels for Marine Production**					
海洋渔业船舶 Marine Fishing Vessels	297 734	707.12			15 622 494
远洋渔船 Ocean-going Fishing Vessels	1 546				915 272
海洋油气船舶 Offshore Oil and Gas Vessels	165	79.00	35.00	8 964	736 593
钻井平台 Drilling Vessels	35	29.00	9.00	4 852	183 152
物探船 Physical Exploration Vessels	8	1.70	0.60	394	24 601
其 他 Others	122	47.00	25.00	3 718	528 840
海洋运输船舶 Maritime Transport Vessels	12 686	6 893	10 605	178 589	29 068 795
二、海洋科研用船 **Vessels for Marine Scientific Research**					
海洋地质勘探船 Marine Geology Survey Vessels	5	1.13	0.36	389	21 165
海洋调查船 Marine Research Vessels	9	3.32	1.46	508	43 531
中国科学院 Chinese Acadamy of Sciences	5	0.72	0.11	190	18 519
国家海洋局 State Oceanic Administration	4	2.60	1.35	318	25 012
三、海洋执法用船 **Vessels for Marine Law Enforcement**					
海监船 Marine Monitoring Vessels	50				

4-9 沿海规模以上港口生产用码头泊位
Berths for Productive Use at above Designed Size Seaports

单位：米, 个 (m, number)

港 口 Seaport		码头长度 Length of Quay Line	泊位个数 Number of Berths	#万吨级 10 000 Tonnage Class
合 计	**Total**	**595 483**	**4 661**	**1 293**
丹 东	Dandong	4 814	33	16
大 连	Dalian	33 686	200	78
营 口	Yingkou	13 244	63	38
锦 州	Jinzhou	5 375	20	18
秦皇岛	Qinhuangdao	14 750	66	42
黄 骅	Huanghua	4 204	20	14
唐 山	Tangshan	10 901	44	41
天 津	Tianjin	30 567	140	95
烟 台	Yantai	14 166	75	46
威 海	Weihai	3 225	15	10
青 岛	Qingdao	19 500	75	59
日 照	Rizhao	11 625	47	40
上 海	Shanghai	72 537	602	150
连云港	Lianyungang	10 158	53	38
宁波-舟山	Ningbo-Zhoushan	6 115	42	19
嘉 兴	Jiaxing	71 668	650	120
台 州	Taizhou	10 474	169	5
温 州	Wenzhou	16 247	234	15
宁 德	ningde	4 902	48	2
福 州	Fuzhou	15 697	122	40

4-9 续表 continued

港　口 Seaport		码头长度 Length of Quay Line	泊位个数 Number of Berths	#万吨级 10 000 Tonnage Class
莆　田	Putian	3 881	42	4
泉　州	Quanzhou	13 366	102	19
厦　门	Xiamen	20 540	110	57
漳　州	Zhangzhou	1 807	19	1
汕　头	Shantou	9 444	86	18
汕　尾	Shanwei	1 580	14	1
惠　州	Huizhou	6 694	34	18
深　圳	Shenzhen	29 971	160	68
虎　门	humen	10 522	91	17
广　州	Guangzhou	42 001	473	60
中　山	Zhongshan	3 447	60	0
珠　海	Zhuhai	12 390	122	16
江　门	Jiangmen	9 196	144	2
阳　江	Yangjiang	1 407	7	6
茂　名	Maoming	2 333	18	8
湛　江	Zhanjiang	15 757	153	31
北部湾	Beibuwan	24 694	217	49
洋　浦	Yangpu	5 647	52	10
海　口	Haikou	5 197	29	14
八　所	Basuo	1 754	10	8

4-10 沿海地区星级饭店基本情况
Star-rated Hotels and Occupancies by Coastal Regions

地　区 Region	饭店数（座） Number of Hotels (unit)	客房数（间） Number of Rooms (unit)	床位数（张） Number of Beds (unit)	客房出租率 （%） Room Occupancy (%)
合　计　Total	**5 378**	**730 180**	**1 234 781**	**56.45**
天　津 Tianjin	99	16 235	27 445	50.68
河　北 Hebei	198	24 258	44 321	57.51
辽　宁 Liaoning	432	52 678	88 920	57.98
上　海 Shanghai	291	63 975	98 909	65.35
江　苏 Jiangsu	702	92 122	154 499	61.24
浙　江 Zhejiang	814	102 979	176 407	62.90
福　建 Fujian	374	48 952	82 095	63.14
山　东 Shandong	895	97 105	171 384	66.66
广　东 Guangdong	1 008	152 891	249 708	57.57
广　西 Guangxi	379	47 826	84 703	62.12
海　南 Hainan	186	31 159	56 390	59.65

4-11 沿海地区旅行社数
Number of Travel Agencies by Coastal Regions

单位：家　　(number)

地 区 Region	旅行社总数 Number of Travel Agencies
合 计 Total	**11 448**
天 津 Tianjin	310
河 北 Hebei	1 148
辽 宁 Liaoning	1 145
上 海 Shanghai	867
江 苏 Jiangsu	1 805
浙 江 Zhejiang	1 639
福 建 Fujian	718
山 东 Shandong	1 842
广 东 Guangdong	1 247
广 西 Guangxi	428
海 南 Hainan	299

主要统计指标解释

1. 海水可养殖面积 指利用滩涂、浅海、港湾进行鱼、虾、蟹、贝、藻等海水经济动植物的人工养殖的水面面积。

2. 浅海养殖 指在可养殖的浅海中进行海水经济动植物养殖。

3. 港湾养殖 指利用港、湾，或在海边、河口附近的滩涂、洼地拦闸筑堤进行海水养殖。

4. 海水养殖面积 是指利用海上、滩涂、陆基进行鱼、甲壳类（虾、蟹）、贝、藻等海水经济动植物的人工养殖的水面面积。在报告期内无论是否全部收获或尚未收获其产品，均应统计在海水养殖面积中。但有些滩涂、水面不投放苗种或投放少量苗种，只进行一般管理的，不统计为养殖面积。

5. 盐田总面积 指盐田占有的全部面积。包括储卤、蒸发、保卤、结晶面积、滩内的沟、壕、池、埝、滩坨等面积及滩外的沟、壕、公路及杂地面积。

6. 生产面积 指直接提供给海盐生产的面积，包括结晶面积、蒸发面积、保卤面积，滩内的沟、壕、池、埝面积及滩坨面积。

7. 年末海盐生产能力 指年末企业生产原盐的全部设备的综合平衡能力。海盐生产露天作业，受天气影响，因而计算生产能力时，成熟滩田按 10 年实际平均单位生产面积产量乘以本年成熟滩田生产面积而得，新滩田按设计能力及滩田成熟程度可能达到的产量计算。

8. 海洋渔业船舶 是指配置机器作为动力的从事海洋渔业生产和辅助渔业生产的船舶。

9. 远洋渔船 按我国远洋渔业项目管理办法在非我国管辖海域（外国专属经济区水域或公海）进行常年或季节性生产的渔船。

10. 泊位个数 是指设有系靠船舶设施，在同一时间内可供靠泊最大吨级船舶的艘数。即可靠泊一艘船舶，则计为一个泊位，余类推。泊位分码头泊位和浮筒泊位。

11. 客房数 指饭店实际可用于接待旅游者的房间数。

12. 床位数 指饭店实际可用于接待旅游者的床位数。

Explanatory Notes on Main Statistical Indicators

1. Marine Cultivatable Area refers to water areas in tidal flat, shallow sea and bay that is used to breed marine economic animals and plants, such as fish, shrimp, crab, shellfish, alga and so on.

2. Shallow Sea Cultivation refers to the breeding of marine economic animals and plants in the cultivatable shallow sea.

3. Harbor Cultivation refers to the marine cultivation conducted in harbors, bays, or the tidal flat or marshes around seaside and bayou by blocking the gate and banking up the dam.

4. Mariculture Area refers to the area of the water surface where seawater economic animals and plants, such as crustacean (shrimp, crab), shellfish and algae, are cultivated at sea, on tidal flat and land. Whether or not all the products in the area have been harvested or the products have not been harvested yet in the period covered by the report, the area is included in the Mariculture Area. But some tidal flats and water surfaces where none or a small amount of the young have been released and only general management is carried out are not included in the Mariculture Area.

5. Total Area of Salt Pans refers to the total area covered by salt pans, including the area for brine storage, evaporation, brine preservation, and crystallization, the area of ditches, moats, pondsand banks within the beach as well as beach mounds, and ditches, moats, highway beyond the beach as well as the area of miscellaneous lands.

6. Area of Salt Pan Production refers to the area directly provided for sea-salt productions, including the area for crystallization, evaporation and brine preservation, the area of ditches, moats, ponds and banks within the beach as well as the area of beach mounds.

7. Year-End Capacity of Crude Salt Production refers to the integrated and balanced capacity of all equipment of the enterprise used for crude salt production at the end of the year. As sea salt production is an open-air operation, which is subject to the effect of weather, the production capacity of a matured salt pan is calculated at the productions of the actual average unit production area in ten years times the production area of the matured salt pan in the current year. The production capacity of new salt pans is calculated at the production that may be reached in the light of the designed capacity and the level of maturity of the salt pan.

8. Marine Fishing Vessels refer to the vessels equipped with machines as motive power and going for marine fishery production and auxiliary fishery production.

9. Deep Sea Fishing Vessels refer to the fishing vessels which carry out production all the year round or seasonally in the non-Chinese jurisdictional, sea areas (foreign EEZ or high sea) according to the China Deep-Sea Fishing Projects Management Measures.

10. Number of Berths refers to the spaces equipped with facilities for docking ships and the number of ships of the maximum tonnage that may dock or anchor in them. A space for a ship to dock is counted as one berth and the rest are reasoned out by analogy. Berths are divided into wharf berths and buoy berths.

11. Number of Rooms refers to the number of guest rooms actually used by the hotels receiving tourists.

12. Number of Beds refers to the number of beds actually used by the hotels receiving tourists.

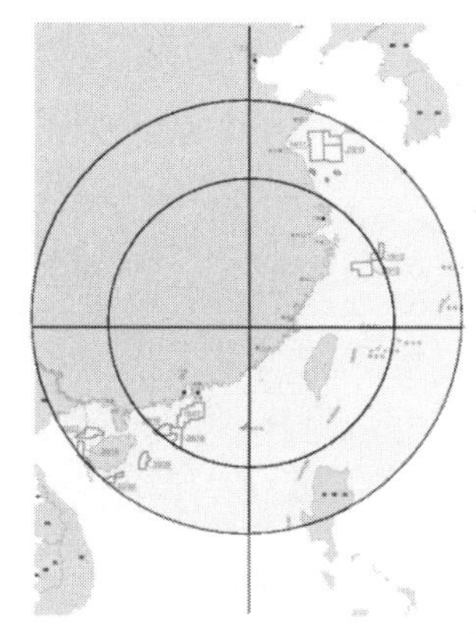

5

涉 海 就 业

Ocean-Related Employment

5-1 全国涉海就业人员情况
National Ocean-Related Employed Personnel

地 区 Region	2001		2009[①]		2010[①]	
	总数（万人） Total Number (10 000 persons)	占地区就业人员比重（%） Proportion in the Regional Employed Personnel (%)	总数（万人） Total Number (10 000 persons)	占地区就业人员比重（%） Proportion in the Regional Employed Personnel (%)	总数（万人） Total Number (10 000 persons)	占地区就业人员比重（%） Proportion in the Regional Employed Personnel (%)
合 计 Total	**2 107.6**	**8.1**	**3 270.6**	**10.1**	**3 350.8**	**10.1**
天 津 Tianjin	106.4	25.9	165.1	32.6	169.2	32.5
河 北 Hebei	58.0	1.7	90.0	2.3	92.2	2.4
辽 宁 Liaoning	196.0	10.7	304.2	13.9	311.6	13.9
上 海 Shanghai	127.5	18.4	197.9	21.3	202.7	21.9
江 苏 Jiangsu	116.9	3.3	181.4	4.0	185.9	3.9
浙 江 Zhejiang	256.4	9.2	397.9	10.4	407.6	10.2
福 建 Fujian	259.7	15.5	403.0	18.6	412.9	18.9
山 东 Shandong	319.9	6.8	496.4	9.1	508.6	9.0
广 东 Guangdong	505.3	12.8	784.1	13.9	803.4	13.9
广 西 Guangxi	68.9	2.7	106.9	3.7	109.5	3.7
海 南 Hainan	80.6	23.7	125.1	29.0	128.1	28.7
其 他[②] Others	12.0		18.6		19.1	

注：① 2009年和2010年为推算数据；② 其他为非沿海地区涉海就业人员数。

Note: ① The data for 2009 and 2010 are the estimated ones;

② Others refer to the number of ocean-related employed persons in the non-coastal regions.

5-2 全国主要海洋产业就业人员情况
Employed Personnel in the Major Marine Industries throughout the Country

单位：万人 (10 000 persons)

海洋产业 Marine Industry	2001	2009	2010
合计 Total	**719.1**	**1 115.0**	**1 142.2**
海洋渔业及相关产业 Marine Fishery and the Related Industries	348.3	540.0	553.2
海洋石油和天然气业 Offshore Oil and Natural Gas Industry	12.4	19.2	19.7
海滨砂矿业 Beach Placer Industry	1.0	1.6	1.6
海洋盐业 Sea Salt Industry	15.0	23.3	23.8
海洋化工业 Marine Chemical Industry	16.1	25.0	25.6
海洋生物医药业 Marine Biomedicine Industry	0.6	0.9	1.0
海洋电力和海水利用业 Marine Electric Power and Seawater Utilization Industry	0.7	1.1	1.1
海洋船舶工业 Marine Shipbuilding Industry	20.6	31.9	32.7
海洋工程建筑业 Marine Engineering Architecture Industry	38.8	60.2	61.6
海洋交通运输业 Maritime Communications and Transportation Industry	50.8	78.8	80.7
滨海旅游业 Coastal Tourism	78.3	121.4	124.4
其他海洋产业 Other Marine Industries	136.5	211.6	216.8

注：2009年和2010年为推算数据。

Note: The data for 2009 and 2010 are the estimated ones.

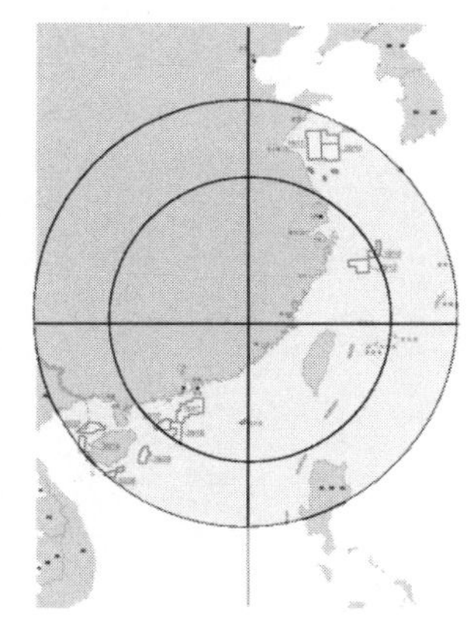

6

海洋科学技术

Marine Science and Technology

6-1 分行业海洋科研机构及人员情况
Marine Scientific Research Institutions and Personnel by Industry

行　业 Industry	机构数（个） Number of Institutions(unit)	从业人员（人） Employed Population(person)
合 计 **Total**	**181**	**35 405**
海洋基础科学研究 **Marine Basic Scientific Research**	**104**	**16 431**
海洋自然科学 Marine Natural Science	55	12 075
海洋社会科学 Marine Social Science	5	1 084
海洋农业科学 Marine Agricultural Science	40	3 173
海洋生物医药 Marine Biomedicine	4	99
海洋工程技术研究 **Marine Engineering Technology Research**	**65**	**17 179**
海洋化学工程技术 Marine Chemical Engineeing Technology	12	6 082
海洋生物工程技术 Marine Bioengineering Technology	2	214
海洋交通运输工程技术 Marine Communications and Transport Technology	14	3 486

6-1 续表 continued

行 业 Industry	机构数（个） Number of Institutions(unit)	从业人员（人） Employed Population(person)
海洋能源开发技术 Marine Energies Development Technology	4	2 697
海洋环境工程技术 Marine Environmental Engineering Technology	11	889
河口水利工程技术 Eustuarine Water Conservancy Engineering Technology	18	2 980
其他海洋工程技术 Other Marine Engineering Technology	4	831
海洋信息服务业 **Marine Information Service**	**9**	**1 104**
其他海洋信息服务 Other Marine Information Services	9	1 104
海洋技术服务业 **Marine Technological Service Industry**	**3**	**691**
其他海洋专业技术服务 Other Marine Professional and Technological Services	2	123
海洋工程管理服务 Marine Engineering Management Service	1	568

注：机构为县级以上科研机构；行业分类参照《海洋及相关产业分类》标准。

Note: The institutions are the scientific research institutions above the county level;The classification of industries follows the standard *Classification of Marine Industries and the Related Industries*.

6-2 分行业海洋科研机构科技活动人员学历构成

Educational Background Composition of the Personnel Engaged in Scientific and Technological Activities in Marine Scientific Research Institutions by Industry

单位：人 (person)

行 业 Industry	科技活动人员 Personnel Engaged in Scientific Activities	研究生 Postgraduate		大学生 Graduate	大专生 College Graduates
			#博士 Doctor		
合 计 Total	**29 676**	**7 699**	**5 418**	**10 277**	**3 751**
海洋基础科学研究 Marine Basic Scientific Research	**14 047**	**3 394**	**3 539**	**4 307**	**1 616**
海洋自然科学 Marine Natural Science	10 468	2 497	3 168	2 855	1 046
海洋社会科学 Marine Social Science	986	334	106	384	126
海洋农业科学 Marine Agricultural Science	2 510	546	263	1 026	433
海洋生物医药 Marine Biomedicine	83	17	2	42	11
海洋工程技术研究 Marine Engineering Technology Research	**14 069**	**3 843**	**1 797**	**5 206**	**1 952**
海洋化学工程技术 Marine Chemical Engineeing Technology	5 017	1 088	572	1 634	911
海洋生物工程技术 Marine Bioengineering Technology	202	63	6	75	25

6-2 续表 continued

行 业 Industry	科技活动人员 Personnel Engaged in Scientific Activities	研究生 Postgraduate	#博士 Doctor	大学生 Graduate	大专生 College Graduates
海洋交通运输工程技术 Marine Communications and Transport Technology	2 762	831	87	1 363	365
海洋能源开发技术 Marine Energies Development Technology	2 138	688	666	473	198
海洋环境工程技术 Marine Environmental Engineering Technology	795	229	40	425	78
河口水利工程技术 Eustuarine Water Conservancy Engineering Technology	2 380	710	365	923	275
其他海洋工程技术 Other Marine Engineering Technology	775	234	61	313	100
海洋信息服务业 Marine Information Service	**883**	**246**	**55**	**420**	**133**
其他海洋信息服务 Other Marine Information Services	883	246	55	420	133
海洋技术服务业 Marine Technological Service Industry	**677**	**216**	**27**	**344**	**50**
其他海洋专业技术服务 Other Marine Professional and Technological Services	109	14	2	70	16
海洋工程管理服务 Marine Engineering Management Service	568	202	25	274	34

6-3 分行业海洋科研机构科技活动人员职称构成
Technical Title Composition of Personnel Engaged in Scientific and Technological Activities in the Marine Scientific Research Institutions by Industry

单位：人 (person)

行 业 Industry	科技活动人员 Personnel Engaged in Scientific Activities	高级职称 Senior	中级职称 Intermediate	初级职称 Primary
合 计 **Total**	**29 676**	**11 079**	**9 559**	**5 902**
海洋基础科学研究 **Marine Basic Scientific Research**	**14 047**	**5 627**	**4 732**	**2 448**
海洋自然科学 Marine Natural Science	10 468	4 444	3 493	1 622
海洋社会科学 Marine Social Science	986	335	339	168
海洋农业科学 Marine Agricultural Science	2 510	827	880	628
海洋生物医药 Marine Biomedicine	83	21	20	30
海洋工程技术研究 **Marine Engineering Technology Research**	**14 069**	**4 946**	**4 360**	**3 174**
海洋化学工程技术 Marine Chemical Engineeing Technology	5 017	1 677	1 606	900
海洋生物工程技术 Marine Bioengineering Technology	202	47	72	54

6-3 续表 continued

行 业 Industry	科技活动人员 Personnel Engaged in Scientific Activities	高级职称 Senior	中级职称 Intermediate	初级职称 Primary
海洋交通运输工程技术 Marine Communications and Transport Technology	2 762	625	749	1 202
海洋能源开发技术 Marine Energies Development Technology	2 138	1 136	572	242
海洋环境工程技术 Marine Environmental Engineering Technology	795	253	342	182
河口水利工程技术 Eustuarine Water Conservancy Engineering Technology	2 380	1 007	798	347
其他海洋工程技术 Other Marine Engineering Technology	775	201	221	247
海洋信息服务业 **Marine Information Service**	**883**	**305**	**271**	**82**
其他海洋信息服务 Other Marine Information Services	883	305	271	82
海洋技术服务业 **Marine Technological Service Industry**	**677**	**201**	**196**	**198**
其他海洋专业技术服务 Other Marine Professional and Technological Services	109	47	31	24
海洋工程管理服务 Marine Engineering Management Service	568	154	165	174

6-4 分行业海洋科研机构经费收入
Routine Fund Receipts of the Marine Scientific Research Institutions by Industry

单位：万元 (10 000 yuan)

行 业 Industry	经费收入总额 Fund Total	经常费 Routine Fund	科技活动借贷款 Loans in the Scientific and Technological Activities	基本建设中政府投资 Government Investment in the Capital Construction
合 计 **Total**	**19 550 823**	**18 664 100**	**6 000**	**880 723**
海洋基础科学研究 **Marine Basic Scientific Research**	**8 799 662**	**8 243 324**	**0**	**556 338**
海洋自然科学 Marine Natural Science	7 280 892	6 851 183	0	429 709
海洋社会科学 Marine Social Science	348 704	308 815	0	39 889
海洋农业科学 Marine Agricultural Science	1 156 775	1 070 602	0	86 173
海洋生物医药 Marine Biomedicine	13 291	12 724	0	567
海洋工程技术研究 **Marine Technological Service Industry**	**9 806 837**	**9 476 452**	**6 000**	**324 385**
海洋化学工程技术 Marine Chemical Engineeing Technology	3 068 774	3 013 574	6 000	49 200
海洋生物工程技术 Marine Bioengineering Technology	174 813	174 813	0	0

6-4 续表 continued

行 业 Industry	经费收入总额 Fund Total	经常费 Routine Fund	科技活动借贷款 Loans in the Scientific and Technological Activities	基本建设中政府投资 Government Investment in the Capital Construction
海洋交通运输工程技术 Marine Communications and Transport Technology	1 843 469	1 658 114	0	185 355
海洋能源开发技术 Marine Energies Development Technology	2 587 972	2 587 972	0	0
海洋环境工程技术 Marine Environmental Engineering Technology	532 832	532 832	0	0
河口水利工程技术 Eustuarine Water Conservancy Engineering Technology	1 219 361	1 129 531	0	89 830
其他海洋工程技术 Other Marine Engineering Technology	379 616	379 616	0	0
海洋信息服务业 **Marine Information Service**	**480 206**	**480 206**	**0**	**0**
其他海洋信息服务 Other Marine Information Services	480 206	480 206	0	0
海洋技术服务业 **Marine Technological Service Industry**	**464 118**	**464 118**	**0**	**0**
其他海洋专业技术服务 Other Marine Professional and Technological Services	158 837	158 837	0	0
海洋工程管理服务 Marine Engineering Management Service	305 281	305 281	0	0

6-5 分行业海洋科研机构科技课题情况
Marine Science and Technology Research Projects of the Research Institutions by Industry

单位：项 (item)

行　业 Industry	课题数 Number of research subjects	基础研究 Basic Research	应用研究 Applied Research	试验发展 Experimental Development	成果应用 Result Application	科技服务 Scientific and Technological Service
合 计 Total	**13 466**	**3 104**	**3 497**	**2 838**	**1 359**	**2 668**
海洋基础科学研究 Marine Basic Scientific Research	**9 906**	**3 052**	**3 157**	**1 687**	**656**	**1 354**
海洋自然科学 Marine Natural Science	7 914	2 872	2 818	1 048	369	807
海洋社会科学 Marine Social Science	425	1	20	120	18	266
海洋农业科学 Marine Agricultural Science	1 536	178	317	496	264	281
海洋生物医药 Marine Biomedicine	31	1	2	23	5	0
海洋工程技术研究 Marine Engineering Technology Research	**3 352**	**52**	**339**	**1 134**	**690**	**1 137**
海洋化学工程技术 Marine Chemical Engineeing Technology	575	6	66	338	130	35
海洋生物工程技术 Marine Bioengineering Technology	24	0	0	10	4	10

6-5 续表 continued

行 业 Industry	课题数 Number of research subjects	基础研究 Basic Research	应用研究 Applied Research	试验发展 Experimental Development	成果应用 Result Application	科技服务 Scientific and Technological Service
海洋交通运输工程技术 Marine Communications and Transport Technology	610	7	24	95	164	320
海洋能源开发技术 Marine Energies Development Technology	655	15	121	309	45	165
海洋环境工程技术 Marine Environmental Engineering Technology	212	0	12	63	33	104
河口水利工程技术 Eustuarine Water Conservancy Engineering	1 207	22	110	290	303	482
其他海洋工程技术 Other Marine Engineering Technology	69	2	6	29	11	21
海洋信息服务业 Marine Information Service	**119**	**0**	**0**	**6**	**0**	**113**
其他海洋信息服务 Other Marine Information Services	119	0	0	6	0	113
海洋技术服务业 Marine Technological Service Industry	**89**	**0**	**1**	**11**	**13**	**64**
其他海洋专业技术服务 Other Marine Professional and Technological Services	14	0	1	11	2	0
海洋工程管理服务 Marine Engineering Management Service	75	0	0	0	11	64

6-6 分行业海洋科研机构科技论著情况
Marine Scientific and Technological Works of the Research Institutions by Industry

行 业 Industry	发表科技论文（篇） Scientific Theses Published (piece)	#国外发表 Published Abroad	出版科技著作（种） Scientific and Technological Works Published (kind)
合 计 Total	**14 296**	**3 874**	**254**
海洋基础科学研究 Marine Basic Scientific Research	**9 750**	**3 423**	**134**
海洋自然科学 Marine Natural Science	7 953	3 243	84
海洋社会科学 Marine Social Science	263	21	14
海洋农业科学 Marine Agricultural Science	1 517	159	36
海洋生物医药 Marine Biomedicine	17		
海洋工程技术研究 Marine Engineering Technology esearch	**4 187**	**417**	**116**
海洋化学工程技术 Marine Chemical Engineeing Technology	605	56	7
海洋生物工程技术 Marine Bioengineering Technology	2		

6-6 续表 continued

行　业 Industry	发表科技论文（篇） Scientific Theses Published (piece)	#国外发表 Published Abroad	出版科技著作（种） Scientific and Technological Works Published (kind)
海洋交通运输工程技术 Marine Communications and Transport Technology	667	52	7
海洋能源开发技术 Marine Energies Development Technology	1 010	117	33
海洋环境工程技术 Marine Environmental Engineering Technology	187	16	6
河口水利工程技术 Eustuarine Water Conservancy Engineering	1 280	171	55
其他海洋工程技术 Other Marine Engineering Technology	436	5	8
海洋信息服务业 Marine Information Service	**275**	**8**	**4**
其他海洋信息服务 Other Marine Information Services	275	8	4
海洋技术服务业 Marine Technological Service Industry	**84**	**26**	
其他海洋专业技术服务 Other Marine Professional and Technological Services	34	21	
海洋工程管理服务 Marine Engineering Management Service	50	5	

6-7 分行业科研机构科技专利情况
Marine Scientific and Technological Patents of the Research Institutions by Industry

单位：件 (number)

行 业 Industry	专利申请受理数 Number of Patent Applications Accepted	#发明专利 Patents for Discoveries	专利授权数 Number of Patents Granted	#发明专利 Patents for Discoveries	拥有发明专利总数 Total Number of Patents for Discoveries
合 计 Total	**3 829**	**3 275**	**1 482**	**988**	**6 750**
海洋基础科学研究 Marine Basic Scientific Research	**1 153**	**913**	**658**	**410**	**1 981**
海洋自然科学 Marine Natural Science	916	747	505	352	1 777
海洋社会科学 Marine Social Science	3	1	3	3	3
海洋农业科学 Marine Agricultural Science	233	164	150	55	192
海洋生物医药 Marine Biomedicine	1	1	0	0	9
海洋工程技术研究 Marine Engineering Technology Research	**2 655**	**2 348**	**815**	**575**	**4 755**
海洋化学工程技术 Marine Chemical Engineeing Technology	2 168	2 086	583	503	4 376
海洋生物工程技术 Marine Bioengineering Technology	0	0	0	0	0

6-7 续表 continued

行 业 Industry	专利申请受理数 Number of Patent Applications Accepted	#发明专利 Patents for Discoveries	专利授权数 Number of Patents Granted	#发明专利 Patents for Discoveries	拥有发明专利总数 Total Number of Patents for Discoveries
海洋交通运输工程技术 Marine Communications and Transport Technology	108	50	52	11	21
海洋能源开发技术 Marine Energies Development Technology	191	107	81	25	167
海洋环境工程技术 Marine Environmental Engineering Technology	27	16	3	1	1
河口水利工程技术 Eustuarine Water Conservancy Engineering Technology	70	30	59	23	100
其他海洋工程技术 Other Marine Engineering Technology	91	59	37	12	90
海洋信息服务业 Marine Information Service	**0**	**0**	**0**	**0**	**0**
其他海洋信息服务 Other Marine Information Services	0	0	0	0	0
海洋技术服务业 Marine Technological Service Industry	**21**	**14**	**9**	**3**	**14**
其他海洋专业技术服务 Other Marine Professional and Technological Services	19	13	5	2	13
海洋工程管理服务 Marine Engineering Management Service	2	1	4	1	1

6-8 分地区海洋科研机构及人员情况
Marine Scientific Research Institutions and Personnel by Regions

地　区 Region	机构数（个） Number of Institutions (unit)	从业人员（人） Employed Population (person)
合　计 Total	**181**	**35 405**
北　京 Beijing	25	12 878
天　津 Tianjin	14	2 467
河　北 Hebei	5	544
辽　宁 Liaoning	17	1 993
上　海 Shanghai	15	3 370
江　苏 Jiangsu	12	3 090
浙　江 Zhejiang	17	1 396
福　建 Fujian	12	1 004
山　东 Shandong	22	3 610
广　东 Guangdong	25	2 795
广　西 Guangxi	9	446
海　南 Hainan	3	197
其　他 Other	5	1 615

6-9 分地区海洋科研机构科技活动人员学历构成

Educational Background Composition of the Personnel Engaged in Scientific and Technological Activities in Marine Scientific Research Institutions by Regions

单位：人 (person)

地 区 Region	科技活动人员 Personnel Engaged in Scientifical Activities	研究生 Postgraduate	#博士 Doctor	大学生 Graduate	大专生 College Graduates
合 计 Total	**29 676**	**7 699**	**5 418**	**10 277**	**3 751**
北 京 Beijing	10 968	3 017	2 859	2 971	1 277
天 津 Tianjin	1 938	563	97	892	227
河 北 Hebei	498	96	21	269	75
辽 宁 Liaoning	1 610	362	90	687	261
上 海 Shanghai	2 919	780	408	1 036	440
江 苏 Jiangsu	2 509	638	252	1 097	298
浙 江 Zhejiang	1 148	319	91	512	155
福 建 Fujian	974	241	77	376	126
山 东 Shandong	2 940	705	564	1 016	410
广 东 Guangdong	2 299	577	549	744	265
广 西 Guangxi	332	52	10	183	62
海 南 Hainan	172	37	4	63	20
其 他 Other	1 369	312	396	431	135

6-10 分地区海洋科研机构科技活动人员职称构成
Technical Title Composition of Personel Engaged in Marine Scientific Research Institutions by Regions

单位：人 (person)

地 区 Region	科技活动人员 Personnel Engaged in Scientifical Activities	高级职称 Senior	中级职称 Intermediate	初级职称 Primary
合 计 Total	**29 676**	**11 079**	**9 559**	**5 902**
北 京 Beijing	10 968	4 744	3 573	1 690
天 津 Tianjin	1 938	625	658	382
河 北 Hebei	498	196	121	33
辽 宁 Liaoning	1 610	566	443	238
上 海 Shanghai	2 919	958	992	648
江 苏 Jiangsu	2 509	772	602	914
浙 江 Zhejiang	1 148	450	399	174
福 建 Fujian	974	270	311	235
山 东 Shandong	2 940	1 049	1 065	620
广 东 Guangdong	2 299	864	796	472
广 西 Guangxi	332	58	123	126
海 南 Hainan	172	17	22	49
其 他 Other	1 369	510	454	321

6-11 分地区海洋科研机构经费收入
Routine Fund Receipts of Marine Scientific Research Institutions by Regions

单位：万元 (10 000 yuan)

地　区 Region	经费收入总额 Fund Total	经常费 Routine Fund	科技活动借贷款 Loans in the Scientific and Technological Activities	基本建设中政府投资 Government Investment in the Capital Construction
合　计 Total	**19 550 823**	**18 664 100**	**6 000**	**880 723**
北　京 Beijing	7 813 133	7 558 238	0	254 895
天　津 Tianjin	1 597 189	1 450 753	0	146 436
河　北 Hebei	110 173	100 384	0	9 789
辽　宁 Liaoning	862 859	862 709	0	150
上　海 Shanghai	2 262 120	2 204 944	0	57 176
江　苏 Jiangsu	1 320 846	1 267 003	0	53 843
浙　江 Zhejiang	856 514	844 929	0	11 585
福　建 Fujian	436 869	433 314	0	3 555
山　东 Shandong	1 889 731	1 718 145	1 000	170 586
广　东 Guangdong	1 532 153	1 378 255	5 000	148 898
广　西 Guangxi	76 308	75 733	0	575
海　南 Hainan	41 810	39 243	0	2 567
其　他 Other	751 118	730 450	0	20 668

6-12 分地区海洋科研机构科技课题情况
Marine Science and Technology Research Projects of the Research Institutions by Regions

单位：项 (item)

地 区 Region	课题数 Number of research subjects	基础研究 Basic Research	应用研究 Applied Research	试验发展 Experimental Development	成果应用 Result Application	科技服务 Scientific and Technological Service
合 计 Total	**13 466**	**3 104**	**3 497**	**2 838**	**1 359**	**2 668**
北 京 Beijing	4 834	1 212	1 087	881	427	1 227
天 津 Tianjin	485	0	42	197	73	173
河 北 Hebei	51	1	20	1	13	16
辽 宁 Liaoning	257	0	21	124	86	26
上 海 Shanghai	1 088	127	321	312	97	231
江 苏 Jiangsu	1 616	82	495	509	271	259
浙 江 Zhejiang	393	80	58	74	57	124
福 建 Fujian	621	259	171	63	81	47
山 东 Shandong	1 358	376	498	272	88	124
广 东 Guangdong	1 678	438	528	235	86	391
广 西 Guangxi	111	10	27	41	20	13
海 南 Hainan	56	0	0	8	47	1
其 他 Other	918	519	229	121	13	36

6-13 分地区海洋科研机构科技论著情况
Marine Scientific and Technological Works of the Research Institutions by Regions

地 区 Region	发表科技论文（篇） Scientific Theses Published (piece)	#国外发表 Published Abroad	出版科技著作（种） Scientific and Technological Works Published (kind)
合 计 **Total**	**14 296**	**3 874**	**254**
北 京 Beijing	5 626	1 724	147
天 津 Tianjin	668	74	13
河 北 Hebei	90	30	1
辽 宁 Liaoning	316	35	2
上 海 Shanghai	1 032	252	7
江 苏 Jiangsu	1 070	241	18
浙 江 Zhejiang	452	76	5
福 建 Fujian	349	105	13
山 东 Shandong	1 651	522	25
广 东 Guangdong	1 685	420	12
广 西 Guangxi	105	4	0
海 南 Hainan	36	15	0
其 他 Other	1 216	376	11

6-14 分地区海洋科研机构科技专利情况
Marine Scientific and Technological Patents of the Research Institutions by Regions

单位：件 (number)

地 区 Region	专利申请受理数 Number of Patent Applications Accepted	#发明专利 Patents for Discoveries	专利授权数 Number of Patents Granted	#发明专利 Patents for Discoveries	拥有发明专利总数 Total Number of Patents for Discoveries
合 计 Total	**3 829**	**3 275**	**1 482**	**988**	**6 750**
北 京 Beijing	1 714	1 553	615	466	3 487
天 津 Tianjin	86	41	43	24	74
河 北 Hebei	0	0	1	1	7
辽 宁 Liaoning	502	448	117	92	930
上 海 Shanghai	725	651	310	186	1 052
江 苏 Jiangsu	136	61	74	30	71
浙 江 Zhejiang	42	21	25	4	36
福 建 Fujian	31	26	10	5	72
山 东 Shandong	267	232	127	94	254
广 东 Guangdong	229	180	115	70	580
广 西 Guangxi	6	6	0	0	0
海 南 Hainan	3	2	1	0	0
其 他 Other	88	54	44	16	187

主要统计指标解释

1. 海洋科研机构 指有明确的研究方向和任务，有一定水平的学术带头人和一定数量、质量的研究人员，有开展研究工作的基本条件，长期有组织地从事海洋研究与开发活动的机构。

2. 从业人员 指由本机构年末直接组织安排工作并支付工资的各类人员总数。包括固定职工、国家有编制的合同制职工、招聘人员和返聘的离退休人员。不包括离退休人员、停薪留职人员。

3. 从事科技活动人员 指从业人员中的科技管理人员、课题活动人员和科技服务人员。

4. 高级职称 指研究员、副研究员；教授、副教授；高级工程师；高级农艺师；正、副主任医（药、护、技)师；高级实验师；高级统计师；高级经济师；高级会计师；编审(正、副编审)；译审(正、副译审)、高级(主任)记者；正、副研究馆员等。

5. 中级职称 指助理研究员；讲师；工程师；农艺师；主治医(药、护、技)师；实验师；统计师；经济师；会计师；编辑；翻译；记者；馆员等。

6. 初级职称 指研究实习员；助教；助理工程师、技术员；助理农艺师、农业技术员；医(药、护、技)师、医(药、护、技)士；助理实验师、实验员；助理统计师、统计员；助理经济师；助理会计师、会计员；助理编辑、见习编辑；助理翻译；助理记者；助理馆员、管理员等。

7. 科技经费筹集额 指从各种渠道筹集到的计划用于本单位科技活动的经费，不论来源渠道如何。

8. 政府资金 指由各级政府部门直接拨款或企事业单位利用政府资金委托本机构从事科学技术活动所获得的收入。

9. 生产经营活动收入 指本机构在科研、技术等专业业务活动以外开展非独立核算的经营活动取得的收入，包括产品（商品）销售收入、经营服务收入、工程承包收入、租赁收入和其他经营收入。

10. 其他收入 指开展科技活动与生产经营活动以外的各项活动的收入，包括：用于离退休人员的政府拨款。

11. 非科技活动借贷款 指本机构为开展非科技活动从各种渠道获得的各类借、贷款。不论偿还形式、期限和数额如何，均按当年获得的借、贷款额填报。不包括基本建设贷款。

12. 基础研究 为获得新知识而进行的独创性研究。其目的是揭示观察到的现象和事实的基本原理和规律，而不以任何特定的实际应用为目的。

13. 应用研究 为获得新的科学技术知识而进行的独创性研究。它主要针对某一特定的实际应用目的。应用研究通常是为了确定基础研究成果或知识的可能的用途，或是为达到某一具体的、预定的实际目的确定新的方法(原理性)或途径。

14. 试验发展 利用从研究或实际经验获得的知识，为生产新的材料、产品和装置，建立新的工艺和系统，以及对已生产或建立的上述各项进行实质性的改进而进行的系统性工作。

15. 成果应用 为解决 R&D 活动阶段产生的新产品、新装置、新工艺、新技术、新方法、新系统和服务等能投入生产或在实际应用中所存在的技术问题而进行的系统性活动。它不具有创新成分。此类活动包括为达到生产目的而进行的定型设计和试制以及为扩大新产品的生产规模和探索新方法、新技术、新工艺等的应用领域而进行的适应性试验。

16. 科技服务 与科学研究与实验发展有关，并有助于科学技术知识的产生、传播和应用的活动。包括为扩大科技成果的使用范围而进行的示范性推广工作；为用户提供科技信息和文献服务的系统性工作；为用户提供可行性报告、技术方案、建议及进行技术论证等技术咨询工作；自然、生物现象的日常观测、监测，资源的考查和勘探；有关社会、人文、经济现象的通用资料的收集，如统计、市场调查等以及这些资料的常规分析与整理；为社会和公众提供的测试、标准化、计量、计算、质量控制和专利服务，不包括工商企业为进行正常生产而开展的上述活动。

17. 生产性活动 由于业务特殊的工艺设备条件，或掌握某种技术专长或诀窍，所进行的小量非常规生产。

18. 科技论文 在全国性学报或学术刊物上、省部属大专院校对外正式发行的学报或学术刊物上发表的论文以及向国外发表的论文。

19. 科技著作 经过正式出版部门编印出版的科技专著、大专院校教科书、科普著作。

20. 专利申请受理数 当年本单位向专利管理部门提出申请并被受理的职务专利申请件数。

21. 专利授权数 当年由专利管理部门授予本单位专利权的职务专利件数。

Explanatory Notes on Main Statistical Indicators

1. Marine Scientific Research Institution refers to the institution which has definite research orientations and tasks, high-level academic leading personnel and fair-sized, qualified research personnel, and basic conditions for research work and which is engaged for a long time in the marine research and development activities in an organized way.

2. Employees refer to the total number of personnel of various kinds employed and paid by the institution at the end of the year, including fixed employees, contract workers of staff belonging to the state authorized staff , recruited personnel, reemployed retired personnel, but not including the retired and the personnel on leave with pay suspension.

3. Personnel Engaged in Scientific and Technological Activities refers to the personnel for scientific and Technological management, personnel engaged in the activities of research topics and scientific and technological service personnel.

4. Senior Technical Title refer to research scientist, associate research scientist; professor, associate professor; senior engineer; senior agronomist; professor-rank and associate professor-rank doctor (pharmacists, nurses and technicians); senior laboratory technician; senior statisticians; senior economic engineer; chief accountant; senior editor (professor and associate professor ranks); senior translator (professor and associate professor ranks); senior journalist; research librarian (professor and associate professor ranks), etc.

5. Intermediate Technical Title refer to assistant research scientist; lecturer; engineer; agronomist; lecturer-rank doctor (pharmacist, nurse and technician); laboratory technician; lecturer-rank statistician;

economic engineer; accountant; editor; translator; journalist; librarian, etc.

6. Primary Technical Title refers to trainee researcher; assistant; assistant engineer, technician, assistant agronomist, agricultural technician; assistant-rank doctor (pharmacist, nurse and technician); assistant laboratory technician; assistant statistician; assistant economic engineer, assistant accountant; assistant editor, editor on probation; assistant translator; assistant journalist; assistant research librarian, librarian ,etc.

7. Amount of Scientific and Technological Funds Raised refers to the funds raised through all channels planned to be used as funds for the scientific and technological activities in the institution regardless of their source and channels.

8. Funds from government refers to the direct appropriations by the government departments at all levels or the earnings from conducting scientific and technological activities entrusted to the institution by enterprises as institutions by earning the funds from government.

9. Earnings from Production as Business Activities refer to the incomes obtained from the non-independent accounting business activities carried out by the institution beyond the scientific research, technological and professional activities, including these from sale of products(goods), business and service, contracted projects, leasing and other business.

10. Other Incomes refer to those from the various activities conducted other than scientific and technological activities and production and business activities, including the government appropriations for retired personnel.

11. Loan for Non-scientific as Technological Activities refers to the various types loan obtained by the institutions through all channels for carrying out non-scientific and technological activities, not including the load for capital construction. The loan, irrespective of its form of reimbursement, term and amount is filled in a form and submitted to the authorities as the amount acquired in the current year.

12. Basic Research refers to the original research to acquire new knowledge. It is aimed at revealing the basic principles and laws of the phenomena and facts observed, but not at any specific practical applications.

13. Applied Research refers to the original research to acquire new scientific and technological knowledge. It mainly serves the purpose of a particular practical application. The purpose of applied research is usually to define the potential uses of the research finds or knowledge obtained from basic research or to identify new methods (principles) or ways to reach a specific and predetermined goal.

14. Experimental Development refers to the systematic work carried out to establish new technologies and systems for producing new materials, products and equipment by using the knowledge obtained from research or practical experience, or to make substantial improvement of the above-mentioned which have been produced or established.

15. Result Application refers to the systematic activities carried out to solve the technical problems that might crop up in the production or practical application of the new products, devices, technologies, techniques, methods, systems and service occurring in the course of R&D activities.

They do not bring forth new ideas. Such activities include the finalizing design and trial-production for the purpose of production as well as the adaptive tests to expand the production scale of new products and the application areas of new methods, techniques and technologies.

16. Scientific and Technological Service refers to the activities that are associated with the scientific research and experimental development, and contribute to the generation, dissemination and application of scientific and technological knowledge, which include the demonstrative work of popularization to enlarge the use scope of scientific and technological achievements; the systematic work of providing the users with scientific and technological information and literature service; the technical consultation work of providing users with feasibility reports, technical schemes and recommendations and carrying out technical demonstration; routine observation and monitoring of natural and biological phenomena, and the survey and exploration of resources; collection of universal data on the appropriate social, cultural and economic phenomena, such as statistics and market survey, as well as the routine analysis and sorting-out of these data; the provision for the society and the public of such service as testing, standardization, computation, quality control and patent, but not including the type of the above-mentioned activities carried out by industrial and commercial enterprises for the purpose of normal production.

17. Productive Activity refers to the small-scale and non-conventional productions due to the presence of special technologies and equipment or mastery of a particular technical expertise or secret of success.

18. Scientific Treatises refer to the theses published in the national journals or academic publications, those officially issued journals or academic publications by universities and colleges under provinces or ministries as well as theses published abroad.

19. Scientific and Technological Works refer to the scientific and technological monographs, textbooks for universities and colleges and popular science books published by the official publishing houses.

20. Number of Patents Applied and Accepted refers to the number of professional patent applications of the unit to the patent administrative department and accepted by it in the year.

21. Number of Patents Granted refers to the number of the professional patents granted to the unit by the patent administrative department in the year.

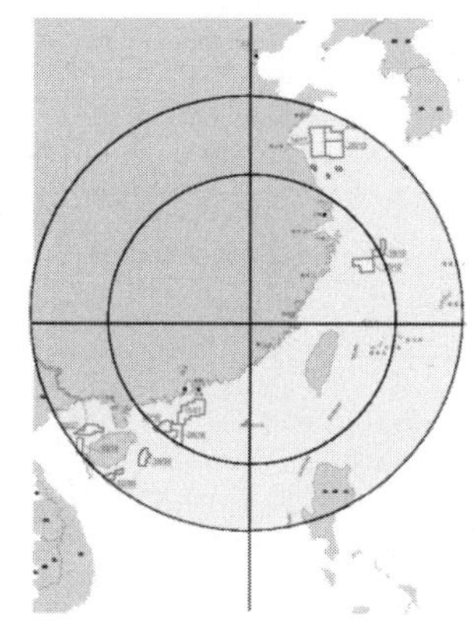

7
海 洋 教 育
Marine Education

7-1 全国各海洋专业博士研究生情况
Doctoral Students from Marine Specialities

专业 Speciality	专业点数（个） Number of Speciality Agencies (unit)	学生数（人） Students (person)			
		毕业生 Graduated	招生 Entrants	在校生 Enrollment	毕业班学生 Graduation
合计 Total	**124**	**679**	**818**	**3 445**	**1 678**
物理海洋学 Physical Oceanography	5	38	58	206	90
海洋化学 Marine Chemistry	5	29	26	124	65
海洋生物学 Marine Biology	9	81	86	282	126
海洋地质 Marine Geology	8	47	55	222	126
海洋科学类新专业 New Speciality of Marine Science	3	67	73	282	145
港口海岸及近海工程 Coastal Harbour and Offshore Engineering	10	53	69	283	110
船舶与海洋结构物设计制造 Ship and Marine Structures Design and Manufacture	10	53	52	356	177
轮机工程 Turbine Engineering	9	29	50	250	93

7-1 续表 continued

专 业 Speciality	专业点数（个） Number of Speciality Agencies (unit)	学生数（人） Students (person)			
		毕业生 Graduated	招 生 Entrants	在校生 Enrollment	毕业班学生 Graduation
水声工程 Hydroacoustic Engineering	7	27	23	148	89
船舶与海洋工程新专业 New Speciality of Shipand Marine Engineering	2	0	4	20	13
捕捞学 Science of Fishing	2	0	3	13	6
航空、航天与航海医学 Aeronautical, Aerospace and Nautical Medicine	0	0	0	0	0
水产品加工及贮藏工程 Aquatic Product Sprocessing and Storing Engineering	8	8	11	44	26
水产新专业 New Specialities of Aquaculture	3	16	8	51	33
水产养殖 Aquaculture	7	33	63	198	86
水生生物学 Hydrobiology	17	71	104	361	158
水文学及水资源 Hydrology and Water Resource	16	121	117	544	305
渔业资源 Fishery Resource	3	6	16	61	30

7-2 全国各海洋专业硕士研究生情况
Postgraduate Students from Marine Specialities

专业 Speciality	专业点数（个） Number of Speciality Agencies (unit)	学生数（人）Students (person)			
		毕业生 Graduated	招生 Entrants	在校生 Enrollment	毕业班学生 Graduation
合计 Total	**291**	**2 915**	**3 569**	**10 363**	**3 392**
物理海洋学 Physical Oceanography	10	64	114	316	101
海洋化学 Marine Chemistry	14	88	103	329	116
海洋生物学 Marine Biology	24	259	409	1 172	344
海洋地质 Marine Geology	15	102	153	395	116
海洋科学类新专业 New Speciality of Marine Science	3	40	71	191	54
港口海岸及近海工程 Coastal Harbour and Offshore Engineering	22	265	319	902	315
船舶与海洋结构物设计制造 Ship and Marine Structures Design and Manufacture	19	325	369	1 220	438
轮机工程 Turbine Engineering	15	268	292	813	306

7-2 续表 continued

专 业 Speciality	专业点数（个） Number of Speciality Agencies (unit)	学生数（人）Students (person)			
		毕业生 Graduated	招 生 Entrants	在校生 Enrollment	毕业班学生 Graduation
水声工程 Hydroacoustic Engineering	11	129	138	444	167
船舶与海洋工程新专业 New Specialities in Shipping and Marine Engineering	5	18	14	61	33
捕捞学 Science of Fishing	4	26	28	90	27
航空、航天与航海医学 Aeronautical, Aerospace and Nautical Medicine	3	15	16	45	14
水产品加工及贮藏工程 Aquatic Products Processing and Storing Engineering	18	109	95	267	83
水产新专业 New Specialities of Aquaculture	5	19	16	58	23
水产养殖 Aquaculture	28	413	448	1 290	406
水生生物学 Hydrobiology	37	247	392	1 095	312
水文学及水资源 Hydrology and Water Resource	50	432	507	1 430	460
渔业资源 Fishery Resource	8	96	85	245	77

注：此表为硕士研究生情况。

Note: The table shows the situation of postgraduate students.

7-3 全国普通高等教育各海洋专业本、专科学生情况
Undergraduates and Students from Colleges for Professional Training from Marine Specialities in the Ordinary National Higher Education

专 业 Speciality	专业点数（个） Number of Speciality Agencies (unit)	学生数（人）Students (person)			
		毕业生 Graduated	招 生 Entrants	在校生 Enrollment	毕业班学生 Graduation
合 计 Total	**607**	**44 653**	**50 169**	**164 246**	**49 264**
海洋科学 Marine Science	19	585	778	2 936	717
海洋技术 Marine Technology	15	400	550	2 193	517
海洋管理 Marine Management	4	41	95	318	37
海洋生物资源与环境 Marine Living Resources and Environment	4	51	95	658	137
海洋科学类新专业 New Speciality of Marine Science	1	0	81	166	0
港口海岸及治河工程 Coastal Harbour and River Control Engineering	1	46	0	115	53
港口航道与海岸工程 Harbour Channel and Coastal Engineering	20	1 019	1 376	5 091	1 176
水资源与海洋工程 Water Resource and Ocean Engineering	2	23	0	134	53
航海技术 Nautical Technology	15	1 739	2 261	8 922	1 937
轮机工程 Turbine Engineering	21	2 066	2 791	10 941	2 537

7-3 续表 1 continued

专 业 Speciality	专业点数（个） Number of Speciality Agencies (unit)	学生数（人）Students (person)			
		毕业生 Graduated	招 生 Entrants	在校生 Enrollment	毕业班学生 Graduation
海事管理 Maritime Affairs Manegernent	2	123	128	496	130
船舶与海洋工程 Ship and Marine Engineering	26	2 142	2 787	11 549	2 796
海洋工程类新专业 New Speciality of Marine Engineering	1	0	210	437	0
水产养殖学 Aquaculture	46	2 135	2 795	9 848	2 484
海洋渔业科学与技术 Science and Technology of Marine Fishery	9	279	424	1 447	289
水族科学与技术 Science and Technology of Aquatic Animals	5	144	256	825	143
水产类新专业 New Speciality in Aquatic Products	1	74	37	78	41
报关与国际货运 Customs Clearing and International Freight Transport	161	14 895	13 750	43 568	15 140
船舶工程技术 Ship Engineering	32	4 234	5 648	17 051	5 684
船舶检验 Ship Inspection	3	131	99	384	121
船舶舾装 Ship Equipment and Installations	3	0	77	152	0
船机制造与维修 Ship Engines Manufacturing and Maintenance	5	0	291	633	136

7-3 续表 2 continued

专 业 Speciality	专业点数（个） Number of Speciality Agencies (unit)	学生数（人）Students (person)			
		毕业生 Graduated	招 生 Entrants	在校生 Enrollment	毕业班学生 Graduation
船艇动力管理 Ship and Boat Power Equipment Management	1	77	42	161	40
港口工程技术 Harbour Engineering	8	557	515	2 382	900
港口机械应用技术 Applied Harbour Machinery Technology	2	381	243	702	296
港口物流设备与自动控制 Harbour Logistics Equipment and Automatic Control	17	1 374	1 102	3 841	1 485
港口业务管理 Harbour Business Management					
港口与航运管理 Harbour and Shipping Management	5	325	335	942	286
港口运输类新专业 New Specialities in Harbour Transportation	3	102	280	628	57
国际航运业务管理 International Shipping Business Management	28	1 980	1 883	5 571	1 767
海关管理 Customs Management	5	380	13	466	262
海关国际法律条约与公约 Customs International Legal treaties and Conventions	1	50	0	0	0
航道工程技术 Channel Engineering	1	0	38	80	0
集装箱运输管理 Container Transportation Management	20	1 064	1 120	3 425	1 190
轮机工程技术 Engine Engineering	42	5 184	6 451	18 179	5 748

7-3 续表 3 continued

专 业 Speciality	专业点数（个） Number of Speciality Agencies (unit)	学生数（人）Students (person)			
		毕业生 Graduated	招 生 Entrants	在校生 Enrollment	毕业班学生 Graduation
水产养殖技术 Aquaculture Technology	26	1 001	1 164	3 656	1 242
水产养殖类新专业 New Specialities in Aquaculture	2	107	92	266	98
水环境监测与保护 Water Environmental Monitoring and Protection	7	164	269	761	235
水上运输类新专业 New Specialities in Waterborne Communications	7	0	468	816	99
物流管理 Logistics management	18	1 282	1 091	2 874	900
水生动植物保护 Aquatic Animals and Plants Protection	3	73	66	222	81
水文与水资源类 Hydrology and Water Resources	1	28	0	0	0
水文与水资源类新专业 New Specialities in Hydrology and Water Resources	0	0	0	0	0
水信息技术 Hydrological Information Technology	1	82	38	75	37
水运管理 Water Transport Management	3	98	126	275	95
水政水资源管理 Water Resources Management by Water Administration	1	20	0	0	0
港口航道与治河工程 Harbour, channel and river harnessing engineering	5	156	210	699	277
特种水产养殖 Special Aquaculture	0	0	0	0	0
渔业综合技术 Integrated Fishery Technologies	4	41	94	283	41

7-4 全国成人高等教育各海洋专业学生情况
Students from Marine Specialities in the National Adult Higher Education

专 业 Speciality	专业点数（个） Number of Speciality Agencies (unit)	学生数（人）Students (person)			
		毕业生 Graduated	招 生 Entrants	在校生 Enrollment	毕业班学生 Graduation
合 计 Total	**161**	**10 336**	**11 513**	**28 375**	**10 673**
海洋科学 Marine Science	1	34	2	22	14
水文与水资源工程 Hydrological and Water Resources Engineering	24	733	673	1 637	720
港口航道与海岸工程 Harbour Channel and Coastal Engineering	5	119	25	109	40
港口海岸及治河工程 Harbour Coastal and river-harnessing Engineering	1	51	0	27	22
航海技术 Navigation Technology	32	3 898	4 096	10 193	4 037
轮机工程 Turbine Engineering	33	3 535	3 538	8 449	3 209
海事管理 Maritime Affairs Management	5	130	65	191	79
船舶与海洋工程 Ship and Marine Engineering	24	986	2 570	6 326	2 000
海洋工程类新专业 New Speciality of Marine Engineering	2	0	34	171	84
水产养殖学 Aquaculture	25	575	287	875	334
海洋渔业科学与技术 Science and Technology of Marine Fishery	2	152	105	118	13
水产类新专业 New Speciality in Aquatic Products	5	48	118	188	52
航运管理 Shipping Management	2	75	0	69	69

7-5 全国中等职业教育各海洋专业学生情况
Students from Marine Specialities in the National Secondary Vocational Education

专 业 Speciality	专业点数（个） Number of Speciality Agencies (unit)	学生数（人）Students (person)			
		毕业生 Graduated	招 生 Entrants	在校生 Enrollment	毕业班学生 Graduation
合 计 Total	**484**	**33 269**	**70 594**	**164 831**	**46 195**
海水生态养殖 Seawater Ecological Cultivation	17	567	1 819	3 286	712
航海捕捞 Marine Fishing	7	134	414	898	198
农林牧渔类新专业 New Specialties in Agriculture, Forestry, Animal Husbandry and Fishery	104	8 999	31 275	78 780	12 374
水文与水资源勘测 Hydrological and Water Resources Survey	4	299	484	768	293
风电场机电设备运行与维护 Operation and Maintenance of Electromechanical Equipment in the Wind Power Station	30	497	2 750	5 275	1 287
船舶制造与修理 Ships Building and Repair	58	2 360	2 717	8 796	3 573
船舶机械装置安装与维修 Installation and Maintenance of Ships' Mechanical Equipment	14	416	424	1 103	347
船舶驾驶 Ship Piloting	86	8 883	14 646	30 570	13 156
轮机管理 Engines Management	60	6 405	10 602	21 197	9 664
船舶水手与机工 Ship Sailors and Mechanics	46	2 344	2 652	7 373	2 543
船舶电气技术 Ship Electric Technology	17	355	486	1 404	667
外轮理货 Foreign Ships Freight Forwarding	14	310	592	1 595	371
船舶检验 Ships Test	7	156	148	522	230
港口机械运行与维护 Operation and Maintenance of Harbour Machinery	17	430	560	1 276	354
工程潜水 Engineering Diving	3	1 114	1 025	1 988	426

注：此表不包含技工学校相关数据。
Note: Data related to Technical Schools are not included in the table.

7-6 分地区海洋专业博士研究生情况
Doctoral Students in Marine Specialities by Regions

地 区 Region	专业点数（个） Number of Speciality Agencies (unit)	学生数（人）Students (person)			
		毕业生 Graduated	招 生 Entrants	在校生 Enrollment	毕业班学生 Graduation
合 计 **Total**	**124**	**679**	**818**	**3 445**	**1 678**
北 京 Beijing	6	14	30	109	51
天 津 Tianjin	3	10	0	13	13
辽 宁 Liaoning	5	43	53	324	57
上 海 Shanghai	17	42	49	266	146
江 苏 Jiangsu	13	87	76	380	223
浙 江 Zhejiang	3	0	13	34	10
福 建 Fujian	6	21	29	117	61
山 东 Shandong	19	245	265	1 050	550
广 东 Guangdong	11	61	76	235	89
其 他 Others	41	156	227	917	478

7-7 分地区海洋专业硕士研究生情况
Postgraduate Students in Marine Specialities by Regions

地 区 Region	专业点数（个） Number of Speciality Agencies (unit)	学生数（人）Students (person)			
		毕业生 Graduated	招 生 Entrants	在校生 Enrollment	毕业班学生 Graduation
合 计 Total	**291**	**2 915**	**3 569**	**10 363**	**3 392**
北 京 Beijing	18	104	120	320	88
天 津 Tianjin	8	91	36	95	25
河 北 Hebei	5	22	18	57	20
辽 宁 Liaoning	20	364	399	1 068	392
上 海 Shanghai	27	341	370	1 191	421
江 苏 Jiangsu	22	330	394	1 118	337
浙 江 Zhejiang	20	138	161	487	145
福 建 Fujian	15	115	181	459	121
山 东 Shandong	35	330	520	1 499	464
广 东 Guangdong	24	222	283	807	264
广 西 Guangxi	2	15	23	75	27
海 南 Hainan	2	20	28	79	24
其 他 Others	93	823	1 036	3 108	1 064

7-8 分地区普通高等教育海洋专业本、专科学生情况
Undergraduates and Students from Colleges for Professional Training in the Marine Specialities of Ordinary Higher Education by Regions

地 区 Region	专业点数（个） Number of Speciality Agencies (unit)	学生数（人）Students (person)			
		毕业生 Graduated	招 生 Entrants	在校生 Enrollment	毕业班学生 Graduation
合 计 Total	**607**	**44 653**	**50 169**	**164 246**	**49 264**
北 京 Beijing	3	57	153	627	155
天 津 Tianjin	25	2 158	2 445	7 853	2 483
河 北 Hebei	30	1 551	1 422	4 964	1 642
辽 宁 Liaoning	39	3 522	4 315	14 575	3 809
上 海 Shanghai	51	4 330	4 077	13 487	4 002
江 苏 Jiangsu	68	6 049	5 974	21 293	6 708
浙 江 Zhejiang	50	2 884	2 898	9 029	2 745
福 建 Fujian	41	3 077	3 038	10 085	3 038
山 东 Shandong	82	6 168	8 002	25 369	7 751
广 东 Guangdong	37	1 951	2 807	8 743	2 347
广 西 Guangxi	20	900	1 339	3 836	942
海 南 Hainan	12	1 259	665	2 463	972
其 他 Others	149	10 747	13 034	41 922	12 670

7-9 分地区成人高等教育海洋专业学生情况
Students from Marine Specialities in the Adult Higher Education by Regions

地 区 Region	专业点数（个）Number of Speciality Agencies (unit)	学生数（人）Students (person)			
		毕业生 Graduated	招 生 Entrants	在校生 Enrollment	毕业班学生 Graduation
合 计 Total	**161**	**10 336**	**11 513**	**28 375**	**10 673**
北 京 Beijing	1	30	0	9	0
天 津 Tianjin	9	875	203	1 112	588
河 北 Hebei	8	200	147	556	151
辽 宁 Liaoning	19	2 221	3 178	7 074	3 271
上 海 Shanghai	10	759	782	1 848	459
江 苏 Jiangsu	12	1 244	1 406	3 649	1 182
浙 江 Zhejiang	12	2 302	1 775	4 883	2 587
福 建 Fujian	9	516	543	1 401	419
山 东 Shandong	23	636	947	1 895	406
广 东 Guangdong	9	292	953	2 040	309
广 西 Guangxi	3	0	2	7	5
其 他 Others	46	1 261	1 577	3 901	1 296

7-10 分地区中等职业教育海洋专业学生情况
Students from Marine Specialities in the Secondary Vocational Education by Regions

地 区 Region	专业点数（个） Number of Speciality Agencies (unit)	学生数（人）Students (person)			
		毕业生 Graduated	招 生 Entrants	在校生 Enrollment	毕业班学生 Graduation
合 计 Total	**484**	**33 269**	**70 594**	**164 831**	**46 195**
天 津 Tianjin	7	891	938	1 860	882
河 北 Hebei	33	2 518	5 640	17 376	1 952
辽 宁 Liaoning	45	1 651	3 149	6 310	2 999
上 海 Shanghai	10	321	251	1 353	583
江 苏 Jiangsu	54	1 176	5 161	13 374	3 676
浙 江 Zhejiang	26	1 339	1 739	4 830	1 703
福 建 Fujian	48	3 591	5 066	13 290	5 997
山 东 Shandong	47	8 621	11 591	23 642	9 454
广 东 Guangdong	16	1 019	6 169	18 128	702
广 西 Guangxi	14	261	1 037	2 697	1 049
海 南 Hainan	1	24	83	198	48
其 他 Others	183	11 857	29 770	61 773	17 150

7-11 分地区开设海洋专业高等学校教职工数
Number of Teaching and Administrative Staff in the Universities and Colleges Offering Marine Specialities by Regions

地 区 Region	机构数（个） Number of Institutions (unit)	教职工数（人） Number of Teaching and Administrative Staff (person)	专任教师数（人） Number of Full-Time Teacher (person)
合 计 Total	**327**	**396 618**	**239 655**
北 京 Beijing	3	5 667	3 157
天 津 Tianjin	12	14 928	9 409
河 北 Hebei	19	17 013	9 955
辽 宁 Liaoning	17	15 232	9 155
上 海 Shanghai	17	22 269	11 534
江 苏 Jiangsu	37	57 356	36 966
浙 江 Zhejiang	19	23 723	13 034
福 建 Fujian	16	12 213	8 210
山 东 Shandong	45	51 855	32 122
广 东 Guangdong	16	23 033	14 012
广 西 Guangxi	14	10 660	7 000
海 南 Hainan	6	5 840	3 526
其 他 Others	106	136 829	81 575

主要统计指标解释

海洋专业 指高等教育和中等职业教育所设的与海洋有关的专业。

Explanatory Notes on Main Statistical Indicators

Marine Speciality refers to the ocean-related speciality in high learning and the professional secondary vocational education.

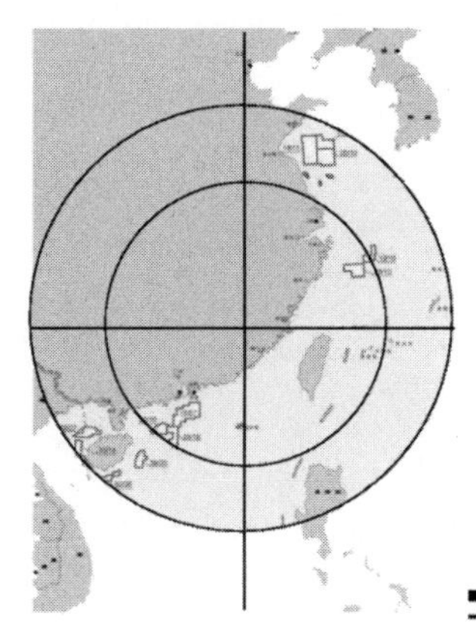

8 海洋环境保护

Marine Environmental Protection

8-1 海区海水水质评价结果
Seawater Quality Assessment Results

项 目 Item	**全国总计 National Total**	渤 海 Bohai Sea	黄 海 Huanghai Sea	东 海 East China Sea	南 海 South China Sea
测站个数 （个） Number of Survey Stations (number)	**1 426**	273	320	431	402
清洁海域面积 （万平方千米） Area of clean sea waters (10 000km^2)		4.43			
较清洁海域面积 （万平方千米） Cleaner Sea Area (10 000km^2)	**7.04**	1.57	1.56	3.28	0.63
轻度污染海域面积 （万平方千米） Lightly-Polluted Sea Area (10 000km^2)	**3.62**	0.87	0.81	1.11	0.83
中度污染海域面积 （万平方千米） Moderately-Polluted Sea Area (10 000km^2)	**2.31**	0.51	0.67	0.93	0.21
严重污染海域面积 （万平方千米） Heavily-Polluted Sea Area (10 000km^2)	**4.80**	0.32	0.65	3.04	0.79
首要超标污染物 Prime Pollutants Exceeding the Set Standard	**无机氮、活性磷酸盐和石油类**	无机氮、活性磷酸盐和石油类	无机氮、活性磷酸盐和石油类	无机氮、活性磷酸盐	无机氮、活性磷酸盐和石油类

8-2 海区废弃物海洋倾倒情况
Ocean Dumping of Wastes by Sea Area

海区 Sea Area	疏浚物 （立方米） Dredged Materials (m^3)	惰性无机地质废料 （立方米） Inert, Inorganic Geologic Wastes (m^3)	人体骨灰 （盒） Human Ashes (case)
合计 **Total**	**199 755 540**	**6 127 000**	**2 509**
渤黄海 Bohai and Huanghai Sea	40 576 000		
东海 East China Sea	136 683 000		2 509
南海 South China Sea	22 496 540	6 127 000	

8-3 海区海洋石油勘探开发污染物排放入海情况
Discharge of Pollutants into the Sea from Offshore Oil Exploration and Exploitation

海 区 Sea Area	生产污水 （万立方米） Sewage from Production (10 000m^3)	泥浆 （立方米） Sludge (m^3)	钻屑 （立方米） Debris from Drilling (m^3)	机舱污水 （立方米） Sewage from Engineroom (m^3)	食品废弃物 （立方米） Food Wastes (m^3)	生活污水 （立方米） Domestic Sewage (m^3)
合 计 Total	**12 165.43**	**55 032.95**	**304 457.50**	**15 328.50**	**1 260.71**	**325 270**
渤 海 Bohai Sea	623.17	11 963.00	288 724.00	4.50	51.36	120 516
黄 海 Huanghai Sea						
东 海 East China Sea	154.65	1 940.95	2 738.50		453.35 *	30 517
南 海 South China Sea	11 387.61	41 129.00	12 995.00	15 324.00	756.00	174 237

注:*单位为吨。

Note: * Unit is ton.

8-4 沿海地区工业废水排放及处理情况
Discharge and Treatment of Industrial Waste Water by Coastal Regions

单位：万吨 (10 000 t)

地 区 Region		工业废水排放总量 Total Volume of Industrial Waste Water Discharged		工业废水中 Industrial Waste Water	
			直接排入海的 Discharged Directly to Sea	符合排放标准的 Up-to-the-Standard	达标率（%） Up-to-the-Standard Rate (%)
总 计 Total		**1 413 764**	**117 954**	**1 369 776**	**96.9**
环渤海经济区 Round-the-Bohai Sea Economic Zone	**合 计 Total**	**413 689**	**35 210**	**403 514**	**97.5**
	辽 宁 Liaoning	71 521	24 919	66 206	92.6
	河 北 Hebei	114 232	415	112 627	98.6
	天 津 Tianjin	19 680	469	19 671	100.0
	山 东 Shandong	208 257	9 406	205 010	98.4
长江三角洲经济区 Yangtze River Delta Economic Zone	**合 计 Total**	**517 882**	**15 082**	**503 789**	**97.3**
	江 苏 Jiangsu	263 760	705	258 622	98.1
	上 海 Shanghai	36 696	2 009	35 973	98.0
	浙 江 Zhejiang	217 426	12 369	209 195	96.2
海峡西岸经济区 Economic Zone on West Side of the Straits	**合 计 Total**	**124 168**	**59 216**	**122 525**	**98.7**
	福 建 Fujian	124 168	59 216	122 525	98.7
珠江三角洲经济区 Zhujiang River Delta Economic Zone	**合 计 Total**	**187 031**	**5 503**	**174 153**	**93.1**
	广 东 Guangdong	187 031	5 503	174 153	93.1
环北部湾经济区 Round-the-Beibu Gulf Economic Zone	**合 计 Total**	**170 993**	**2 944**	**165 795**	**97.0**
	广 西 Guangxi	165 211	756	160 139	96.9
	海 南 Hainan	5 782	2 188	5 656	97.8

8-5 沿海城市工业废水排放及处理情况
Discharge and Treatment of Industrial Waste Water by Coastal Cities

单位：万吨 (10 000 t)

沿海城市 Coastal City	工业废水排放总量 Total Volume of Industrial Waste Water Discharged		工业废水中 Industrial Waste Water	
		直接排入海的 Discharged Directly to Sea	符合排放标准的 Up-to-the-Standard	达标率（%） Up-to-the-Standard Rate (%)
天　津 Tianjin	19 680	469	19 671	100.0
唐　山 Tangshan	18 170	80	17 852	98.2
秦皇岛 Qinhuangdao	5 608	177	5 608	100.0
沧　州 Cangzhou	6 871	92	6 748	98.2
大　连 Dalian	27 421	22 754	26 076	95.1
丹　东 Dandong	4 332	255	3 981	91.9
锦　州 Jinzhou	3 859	87	3 648	94.5
营　口 Yingkou	4 096	902	4 094	100.0
盘　锦 Panjin	2 068	0	1 925	93.1
葫芦岛 Huludao	3 230	914	2 521	78.0
上　海 Shanghai	36 696	2 009	35 973	98.0
南　通 Nantong	15 708	0	15 606	99.4
连云港 Lianyungang	3 538	351	3 473	98.2
盐　城 Yancheng	12 869	1	11 904	92.5

8-5 续表1 continued

沿海城市 Coastal City	工业废水排放总量 Total Volume of Industrial Waste Water Discharged	工业废水中 Industrial Waste Water		
		直接排入海的 Discharged Directly to Sea	符合排放标准的 Up-to-the-Standard	达标率（%） Up-to-the-Standard Rate (%)
杭　州　Hangzhou	80 468	1 240	77 957	96.9
宁　波　Ningbo	18 967	8 328	18 108	95.5
温　州　Wenzhou	17 008	239	16 311	95.9
嘉　兴　Jiaxing	19 812	493	19 533	98.6
绍　兴　Shaoxing	30 230		29 744	98.4
舟　山　Zhoushan	1 493	864	1 443	96.7
台　州　Taizhou	5 709	678	5 153	90.3
福　州　Fuzhou	4 933	155	4 684	95.0
厦　门　Xiamen	4 457	228	4 456	100.0
莆　田　Putian	1 701	524	1 655	97.3
泉　州　Quanzhou	19 544	2 534	19 513	99.8
漳　州　Zhangzhou	62 845	55 308	62 603	99.6
宁　德　Ningde	1 501	408	1 383	92.1
青　岛　Qingdao	10 800	2 714	10 591	98.1
东　营　Dongying	10 559	0	10 559	100.0
烟　台　Yantai	8 386	2 424	8 386	100.0
潍　坊　Weifang	21 496	24	21 210	98.7
威　海　Weihai	2 872	527	2 872	100.0
日　照　Rizhao	9 978	3 693	9 978	100.0
滨　州　Binzhou	15 013	0	14 455	96.3

8-5 续表2 continued

沿海城市 Coastal City	工业废水排放总量 Total Volume of Industrial Waste Water Discharged	工业废水中 Industrial Waste Water		
		直接排入海的 Discharged Directly to Sea	符合排放标准的 Up-to-the-Standard	达标率（%） Up-to-the-Standard Rate (%)
广　州 Guangzhou	23 586	394	22 783	96.6
深　圳 Shenzhen	9 001	563	8 676	96.4
珠　海 Zhuhai	6 125	65	6 004	98.0
汕　头 Shantou	6 150	163	5 219	84.9
江　门 Jiangmen	11 457	491	10 840	94.6
湛　江 Zhanjiang	5 355	206	4 301	80.3
茂　名 Maoming	5 902	994	4 874	82.6
惠　州 Huizhou	6 029	793	5 969	99.0
汕　尾 Shanwei	3 679	2	1 795	48.8
阳　江 Yangjiang	2 790	0	2 152	77.1
东　莞 Dongguan	29 742	1 657	28 639	96.3
中　山 Zhongshan	11 374	0	11 034	97.0
潮　州 Chaozhou	4 237	0	3 762	88.8
揭　阳 Jieyang	3 369	0	3 085	91.6
北　海 Beihai	1 369	62	1 308	95.5
防城港 Fangchenggang	5 889	0	5 640	95.8
钦　州 Qinzhou	4 093	694	4 004	97.8
海　口 Haikou	513	0	513	100.0
三　亚 Sanya	23	0	23	100.0

8-6 沿海地带工业废水排放及处理情况
Discharge and Treatment of Industrial Waste Water by Coastal Counties

单位：万吨 (10 000 t)

地区 Region		工业废水排放总量 Total Volume of Industrial Waste Water Discharged	直接排入海的 Discharged Directly to Sea	工业废水中 Industrial Waste Water 符合排放标准的 Up-to-the-Standard	达标率（%） Up-to-the-Standard Rate (%)
天　津	**Tianjin**	**8 339**	**461**	**8 339**	**100.0**
河　北	**Hebei**	**13 061**	**177**	**12 728**	**97.5**
唐　山	Tangshan	5 629	0	5 319	94.5
秦皇岛	Qinhuangdao	5 000	177	5 000	100.0
沧　州	Cangzhou	2 432		2 409	99.1
辽　宁	**Liaoning**	**30 913**	**20 771**	**28 732**	**92.9**
大　连	Dalian	21 663	18 904	20 405	94.2
丹　东	Dandong	1 140	51	964	84.5
锦　州	Jinzhou	1 779		1 759	98.9
营　口	Yingkou	2 702	902	2 701	100.0
盘　锦	Panjin	505		404	80.0
葫芦岛	Huludao	3 124	914	2 499	80.0
上　海	**Shanghai**	**21 269**	**2 007**	**20 717**	**97.4**
江　苏	**Jiangsu**	**16 497**	**351**	**15 809**	**95.8**
南　通	Nantong	6 437	0	6 364	98.9
连云港	Lianyungang	2 985	351	2 927	98.1
盐　城	Yancheng	7 075	0	6 518	92.1

8-6 续表1 continued

地 区 Region	工业废水排放总量 Total Volume of Industrial Waste Water Discharged	直接排入海的 Discharged Directly to Sea	工业废水中 Industrial Waste Water 符合排放标准的 Up-to-the-Standard	达标率（%） Up-to-the-Standard Rate (%)
浙 江 Zhejiang	**85 127**	**8 727**	**81 428**	**95.7**
杭 州 Hangzhou	22 623	1 240	21 253	93.9
宁 波 Ningbo	13 448	5 222	12 692	94.4
温 州 Wenzhou	12 474	235	11 813	94.7
嘉 兴 Jiaxing	9 391	488	9 330	99.4
绍 兴 Shaoxing	21 453	0	21 055	98.1
舟 山 Zhoushan	1 493	864	1 443	96.7
台 州 Taizhou	4 245	678	3 842	90.5
福 建 Fujian	**85 996**	**59 112**	**85 602**	**99.5**
福 州 Fuzhou	3 186	155	3 026	95.0
厦 门 Xiamen	4 457	228	4 456	100.0
莆 田 Putian	1 701	524	1 655	97.3
泉 州 Quanzhou	17 497	2 534	17 466	99.8
漳 州 Zhangzhou	58 102	55 308	58 004	99.8
宁 德 Ningde	1 053	363	995	94.5
山 东 Shandong	**43 472**	**7 025**	**43 148**	**99.3**
青 岛 Qingdao	4 372	767	4 355	99.6
东 营 Dongying	10 559	0	10 559	100.0
烟 台 Yantai	7 164	2 424	7 164	100.0
潍 坊 Weifang	8 279	0	8 059	97.3
威 海 Weihai	1 881	141	1 881	100.0
日 照 Rizhao	6 797	3 693	6 797	100.0
滨 州 Binzhou	4 420	0	4 333	98.0

8-6 续表2 continued

地 区 Region	工业废水排放总量 Total Volume of Industrial Waste Water Discharged	直接排入海的 Discharged Directly to Sea	工业废水中 Industrial Waste Water 符合排放标准的 Up-to-the-Standard	达标率（%） Up-to-the-Standard Rate (%)
广　东　Guangdong	**96 259**	**4 417**	**88 808**	**92.3**
广　州　Guangzhou	11 094	389	10 665	96.1
深　圳　Shenzhen	7 977	563	7 696	96.5
珠　海　Zhuhai	3 782	30	3 714	98.2
汕　头　Shantou	6 117	163	5 187	84.8
江　门　Jiangmen	8 470	422	8 049	95.0
湛　江　Zhanjiang	4 574	194	3 720	81.3
茂　名　Maoming	3 501	994	2 790	79.7
惠　州　Huizhou	2 181	3	2 148	98.5
汕　尾　Shanwei	3 570	2	1 774	49.7
阳　江　Yangjiang	1 069	0	849	79.4
东　莞　Dongguan	29 742	1 657	28 639	96.3
中　山　Zhongshan	11 374	0	11 034	97.0
潮　州　Chaozhou	1 377	0	1 208	87.7
揭　阳　Jieyang	1 431	0	1 335	93.3
广　西　Guangxi	**6 894**	**753**	**6 563**	**95.2**
北　海　Beihai	992	59	938	94.6
防城港　Fangchenggang	3 793	0	3 561	93.9
钦　州　Qinzhou	2 109	694	2 064	97.9
海　南　Hainan	**512**	**0**	**512**	**100.0**
海　口　Haikou	489	0	489	100.0
三　亚　Sanya	23	0	23	100.0

注：表中数据为沿海地带合计数（表8-7、8-10、8-13同）。

Note: The data in the table are the total numbers for the coastal regions(The same for Tables 8-7,8-10,8-13).

8-7 沿海地区工业固体废物排放、处理及综合利用情况
Discharge, Treatment and Utilization of Industrial Solid Wastes by Coastal Regions

地 区 Region		工业固体废物排放量（吨） Volume of Industrial Solid Wastes Discharged (t)	工业固体废物处置量（万吨） Volume of Industrial Solid Wastes Treated (10 000 t)	工业固体废物综合利用量（万吨） Volume of Industrial Solid Wastes Muti-Utilized (10 000 t)
总 计 Total		**346 845**	**227 786 980**	**740 624 189**
环渤海经济区 Round-the-Bohai Sea Economic Zone	**合 计 Total**	**72 493**	**192 766 183**	**433 255 663**
	辽 宁 Liaoning	27 877	67 673 273	82 097 387
	河 北 Hebei	44 506	120 072 963	179 734 024
	天 津 Tianjin	0	270 248	18 451 172
	山 东 Shandong	110	4 749 699	152 973 080
长江三角洲经济区 Yangtze River Delta Economic Zone	**合 计 Total**	**6 179**	**4 070 242**	**151 603 135**
	江 苏 Jiangsu	0	1 386 916	87 605 889
	上 海 Shanghai	2	938 572	23 669 180
	浙 江 Zhejiang	6 177	1 744 754	40 328 066
海峡西岸经济区 Economic Zone on West Side of the Straits	**合 计 Total**	**35 163**	**11 811 417**	**62 148 942**
	福 建 Fujian	35 163	11 811 417	62 148 942
珠江三角洲经济区 Zhujiang River Delta Economic Zone	**合 计 Total**	**141 609**	**3 504 566**	**49 525 848**
	广 东 Guangdong	141 609	3 504 566	49 525 848
环北部湾经济区 Round-the-Beibu Gulf Economic Zone	**合 计 Total**	**91 401**	**15 634 572**	**44 090 601**
	广 西 Guangxi	91 391	15 630 952	42 306 656
	海 南 Hainan	10	3 620	1 783 945

8-8 沿海城市工业固体废物排放、处理、综合利用情况
Discharge, Treatment and Utilization of Industrial Solid Wastes by Coastal Cities

单位：吨 (t)

沿海城市 Coastal City	工业固体废物排放量 Volume of Industrial Solid Wastes Discharged	工业固体废物处置量 Volume of Industrial Solid Wastes Treated	工业固体废物综合利用量 Volume of Industrial Solid Wastes Muti-Utilized
天　津 Tianjin	0	270 248	18 451 172
唐　山 Tangshan	9 614	16 393 926	85 899 473
秦皇岛 Qinhuangdao	0	4 975 722	8 829 729
沧　州 Cangzhou	0	13 875	3 441 677
大　连 Dalian	7 660	147 540	2 744 620
丹　东 Dandong	0	1 193 088	2 293 403
锦　州 Jinzhou	27	1 250 383	2 099 577
营　口 Yingkou	0	4 154	6 553 067
盘　锦 Panjin	2	36 074	801 329
葫芦岛 Huludao	18 835	975 865	2 619 710
上　海 Shanghai	2	938 572	23 669 180
南　通 Nantong	0	71 242	3 890 953
连云港 Lianyungang	0	3 846	2 827 904
盐　城 Yancheng	0	47 990	1 588 321

8-8 续表1 continued

沿海城市 Coastal City	工业固体废物排放量 Volume of Industrial Solid Wastes Discharged	工业固体废物处置量 Volume of Industrial Solid Wastes Treated	工业固体废物综合利用量 Volume of Industrial Solid Wastes Muti-Utilized
杭　州　Hangzhou	1 800	412 808	6 657 260
宁　波　Ningbo	300	602 799	10 347 507
温　州　Wenzhou	2 733	93 569	2 075 528
嘉　兴　Jiaxing	119	90 102	3 685 940
绍　兴　Shaoxing	0	209 222	3 215 242
舟　山　Zhoushan	0	3 459	776 563
台　州　Taizhou	922	48 382	2 399 739
福　州　Fuzhou	2 820	1 228 484	5 576 096
厦　门　Xiamen	0	97 742	1 179 337
莆　田　Putian	2 437	3 404	498 378
泉　州　Quanzhou	303	185 326	6 484 394
漳　州　Zhangzhou	18	19 064	1 697 898
宁　德　Ningde	1 265	80 148	1 482 939
青　岛　Qingdao	0	111 339	9 074 745
东　营　Dongying	0	129 917	2 225 283
烟　台　Yantai	0	2 470 120	19 080 360
潍　坊　Weifang	10	55 556	7 531 004
威　海　Weihai	0	25 691	2 138 298
日　照　Rizhao	0	917	8 204 942
滨　州　Binzhou	100	31	5 562 400

8-8 续表2 continued

沿海城市 Coastal City	工业固体废物排放量 Volume of Industrial Solid Wastes Discharged	工业固体废物处置量 Volume of Industrial Solid Wastes Treated	工业固体废物综合利用量 Volume of Industrial Solid Wastes Muti-Utilized
广　州 Guangzhou	45	665 819	6 222 214
深　圳 Shenzhen	500	114 524	1 347 378
珠　海 Zhuhai	3 600	45 526	2 700 334
汕　头 Shantou	201	8 641	969 440
江　门 Jiangmen	286	88 201	2 071 355
湛　江 Zhanjiang	9 721	206 257	2 807 688
茂　名 Maoming	361	42 419	2 004 948
惠　州 Huizhou	0	27 425	378 621
汕　尾 Shanwei	21 948	27 773	407 080
阳　江 Yangjiang	7 386	64	1 543 926
东　莞 Dongguan	6 859	150 251	2 972 632
中　山 Zhongshan	391	129 007	905 709
潮　州 Chaozhou	1 200	3 837	845 176
揭　阳 Jieyang	100	2 259	888 729
北　海 Beihai	6 494	439 819	1 260 324
防城港 Fangchenggang	1 400	0	1 587 616
钦　州 Qinzhou	4 760	13 778	787 808
海　口 Haikou	0	1 226	41 172
三　亚 Sanya	0	16	2 903

8-9 沿海地带工业固体废物排放、处理、综合利用情况
Discharge,Treatment and Utilization of Industrial Solid Wastes by Coastal Counties

单位：吨 (t)

地 区 Region	工业固体废物排放量 Volume of Industrial Solid Wastes Discharged	工业固体废物处置量 Volume of Industrial Solid Wastes Treated	工业固体废物综合利用量 Volume of Industrial Solid Wastes Muti-Utilized
天 津 Tianjin	**0**	**170 637**	**6 440 027**
河 北 Hebei	**0**	**34 138**	**15 911 306**
唐 山 Tangshan	0	1 040	9 348 270
秦皇岛 Qinhuangdao	0	27 554	3 713 275
沧 州 Cangzhou	0	5 544	2 849 761
辽 宁 Liaoning	**4 658**	**1 105 833**	**11 113 370**
大 连 Dalian	4 658	114 245	2 391 242
丹 东 Dandong	0	0	904 638
锦 州 Jinzhou	0	11 639	177 767
营 口 Yingkou	0	4 084	6 006 501
盘 锦 Panjin	0	0	57 953
葫芦岛 Huludao	0	975 865	1 575 269
上 海 Shanghai		**618 150**	**20 665 382**
江 苏 Jiangsu	**0**	**40 515**	**5 140 775**
南 通 Nantong	0	15 102	1 692 780
连云港 Lianyungang	0	3 430	2 745 657
盐 城 Yancheng	0	21 983	702 338

8-9 续表1 continued

地 区 Region	工业固体废物排放量 Volume of Industrial Solid Wastes Discharged	工业固体废物处置量 Volume of Industrial Solid Wastes Treated	工业固体废物综合利用量 Volume of Industrial Solid Wastes Muti-Utilized
浙 江 Zhejiang	**4 065**	**1 012 764**	**19 181 966**
杭 州 Hangzhou	0	203 139	2 123 462
宁 波 Ningbo	300	511 284	8 343 490
温 州 Wenzhou	2 733	54 976	1 971 944
嘉 兴 Jiaxing	110	59 345	2 206 156
绍 兴 Shaoxing	0	136 403	1 585 319
舟 山 Zhoushan	0	3 459	776 563
台 州 Taizhou	922	44 158	2 175 032
福 建 Fujian	**3 957**	**1 321 606**	**12 576 177**
福 州 Fuzhou	1 000	1 087 650	4 872 097
厦 门 Xiamen	0	97 742	1 179 337
莆 田 Putian	2 437	3 404	498 378
泉 州 Quanzhou	0	119 830	3 881 556
漳 州 Zhangzhou	0	10 796	1 203 360
宁 德 Ningde	520	2 184	941 449
山 东 Shandong	**0**	**2 599 218**	**36 060 290**
青 岛 Qingdao	0	5 163	2 038 930
东 营 Dongying	0	129 917	2 225 283
烟 台 Yantai	0	2 405 165	18 570 249
潍 坊 Weifang	0	36 606	2 657 723
威 海 Weihai	0	21 684	1 629 415
日 照 Rizhao	0	652	7 722 174
滨 州 Binzhou	0	31	1 216 516

8-9 续表2 continued

地 区 Region	工业固体废物排放量 Volume of Industrial Solid Wastes Discharged	工业固体废物处置量 Volume of Industrial Solid Wastes Treated	工业固体废物综合利用量 Volume of Industrial Solid Wastes Muti-Utilized
广 东 Guangdong	**41 802**	**1 006 256**	**18 327 239**
广 州 Guangzhou	0	242 232	3 758 307
深 圳 Shenzhen	500	106 359	1 328 443
珠 海 Zhuhai	1 900	27 752	254 766
汕 头 Shantou	201	8 636	969 440
江 门 Jiangmen	286	83 939	1 854 752
湛 江 Zhanjiang	7 421	179 424	2 672 686
茂 名 Maoming	361	42 389	1 655 363
惠 州 Huizhou	0	6 251	56 744
汕 尾 Shanwei	16 788	27 773	406 380
阳 江 Yangjiang	6 995	64	663 307
东 莞 Dongguan	6 859	150 251	2 972 632
中 山 Zhongshan	391	129 006	905 709
潮 州 Chaozhou	100	244	13 894
揭 阳 Jieyang	0	1 936	814 816
广 西 Guangxi	**2 969**	**453 596**	**1 999 863**
北 海 Beihai	1 569	439 819	647 211
防城港 Fangchenggang	1 400	0	851 932
钦 州 Qinzhou	0	13 777	500 720
海 南 Hainan	**0**	**1 240**	**22 058**
海 口 Haikou	0	1 224	19 155
三 亚 Sanya	0	16	2 903

8-10 沿海地区污染治理项目情况
Pollution Control Projects by Coastal Regions

单位：个 (number)

地 区 Region		当年安排施工项目 Arranged for Construction in the Year		当年竣工项目 Completed in the Year	
		治理废水 Treatment of Waste Water	治理固体废物 Treatment of Solid Wastes	治理废水 Treatment of Waste Water	治理固体废物 Treatment of Solid Wastes
总 计 Total		**1 832**	**167**	**1 662**	**157**
环渤海经济区 Round-the-Bohai Sea Economic Zone	**合 计 Total**	**317**	**71**	**269**	**70**
	辽 宁 Liaoning	34	11	24	11
	河 北 Hebei	36	3	34	3
	天 津 Tianjin	45	9	34	9
	山 东 Shandong	202	48	177	47
长江三角洲经济区 Yangtze River Delta Economic Zone	**合 计 Total**	**696**	**32**	**634**	**31**
	江 苏 Jiangsu	185	13	174	12
	上 海 Shanghai	65	4	56	4
	浙 江 Zhejiang	446	15	404	15
海峡西岸经济区 Economic Zone on West Side of the Straits	**合 计 Total**	**375**	**30**	**349**	**24**
	福 建 Fujian	375	30	349	24
珠江三角洲经济区 Zhujiang River Delta Economic Zone	**合 计 Total**	**322**	**11**	**296**	**11**
	广 东 Guangdong	322	11	296	11
环北部湾经济区 Round-the-Beibu Gulf Economic Zone	**合 计 Total**	**122**	**23**	**114**	**21**
	广 西 Guangxi	109	22	104	20
	海 南 Hainan	13	1	10	1

8-11 沿海城市污染治理项目情况
Pollution Control Projects by Coastal Cities

单位：个 (number)

沿海城市 Coastal City	当年安排施工项目 Arranged for Construction in the Year		当年竣工项目 Completed in the Year	
	治理废水 Treatment of Waste Water	治理固体废物 Treatment of Solid Wastes	治理废水 Treament of Waste Water	治理固体废物 Treatment of Solid Wastes
天　津　Tianjin	45	9	34	9
唐　山　Tangshan	3	0	3	0
秦皇岛　Qinhuangdao	0	0	0	0
沧　州　Cangzhou	1	1	1	1
大　连　Dalian	12	0	7	0
丹　东　Dandong				
锦　州　Jinzhou	4	2	4	2
营　口　Yingkou	0	0	0	0
盘　锦　Panjin	0	7	0	7
葫芦岛　Huludao	3	0	4	0
上　海　Shanghai	65	4	56	4
南　通　Nantong	41	1	41	1
连云港　Lianyungang	12	1	12	1
盐　城　Yancheng	1	0	1	0

8-11 续表1 continued

沿海城市 Coastal City	当年安排施工项目 Arranged for Construction in the Year		当年竣工项目 Completed in the Year	
	治理废水 Treatment of Waste Water	治理固体废物 Treatment of Solid Wastes	治理废水 Treament of Waste Water	治理固体废物 Treatment of Solid Wastes
杭 州 Hangzhou	223	0	203	0
宁 波 Ningbo	17	0	14	0
温 州 Wenzhou	19	0	19	0
嘉 兴 Jiaxing	16	0	12	0
绍 兴 Shaoxing	67	5	58	5
舟 山 Zhoushan	1	3	1	3
台 州 Taizhou	37	2	35	2
福 州 Fuzhou	27	3	26	3
厦 门 Xiamen	32	2	28	2
莆 田 Putian				
泉 州 Quanzhou	213	22	193	16
漳 州 Zhangzhou	26	0	25	0
宁 德 Ningde	1	0	1	0
青 岛 Qingdao	14	0	11	0
东 营 Dongying	14	5	11	5
烟 台 Yantai	13	37	9	37
潍 坊 Weifang	21	0	19	0
威 海 Weihai	10	0	9	0
日 照 Rizhao	7	2	6	2
滨 州 Binzhou	4	0	4	0

8-11 续表2 continued

沿海城市 Coastal City	当年安排施工项目 Arranged for Construction in the Year		当年竣工项目 Completed in the Year	
	治理废水 Treatment of Waste Water	治理固体废物 Treatment of Solid Wastes	治理废水 Treament of Waste Water	治理固体废物 Treatment of Solid Wastes
广　州　Guangzhou	19	1	17	1
深　圳　Shenzhen	54	2	54	2
珠　海　Zhuhai	19	2	19	2
汕　头　Shantou	43	0	41	0
江　门　Jiangmen	13	0	6	0
湛　江　Zhanjiang	6	0	5	0
茂　名　Maoming	1	0	0	0
惠　州　Huizhou	46	0	45	0
汕　尾　Shanwei				
阳　江　Yangjiang	7	0	7	0
东　莞　Dongguan				
中　山　Zhongshan	10	0	10	0
潮　州　Chaozhou	1	0	1	0
揭　阳　Jieyang	4	0	4	0
北　海　Beihai	3	0	3	0
防城港　Fangchenggang	2	0	2	0
钦　州　Qinzhou	10	2	10	2
海　口　Haikou				
三　亚　Sanya				

8-12 沿海地带污染治理项目情况
Pollution Control Projects by Coastal Counties

单位：个 (number)

地 区 Region	当年安排施工项目 Arranged for Construction in the Year		当年竣工项目 Completed in the Year	
	治理废水 Treatment of Waste Water	治理固体废物 Treatment of Solid Wastes	治理废水 Treament of Waste Water	治理固体废物 Treatment of Solid Wastes
天 津 Tianjin	**6**	**2**	**5**	**2**
河 北 Hebei	**1**	**1**	**1**	**1**
唐 山 Tangshan				
秦皇岛 Qinhuangdao				
沧 州 Cangzhou	1	1	1	1
辽 宁 Liaoning	**14**	**2**	**11**	**2**
大 连 Dalian	10		6	
丹 东 Dandong				
锦 州 Jinzhou	1	2	1	2
营 口 Yingkou				
盘 锦 Panjin				
葫芦岛 Huludao	3		4	
上 海 Shanghai	**50**	**4**	**44**	**4**
江 苏 Jiangsu	**24**	**1**	**24**	**1**
南 通 Nantong	12		12	
连云港 Lianyungang	12	1	12	1
盐 城 Yancheng				

8-12 续表1 continued

地 区 Region	当年安排施工项目 Arranged for Construction in the Year		当年竣工项目 Completed in the Year	
	治理废水 Treatment of Waste Water	治理固体废物 Treatment of Solid Wastes	治理废水 Treament of Waste Water	治理固体废物 Treatment of Solid Wastes
浙 江 Zhejiang	**84**	**7**	**74**	**7**
杭 州 Hangzhou	11		11	
宁 波 Ningbo	13		10	
温 州 Wenzhou	19		19	
嘉 兴 Jiaxing			1	
绍 兴 Shaoxing	29	4	21	4
舟 山 Zhoushan	1	3	1	3
台 州 Taizhou	11		11	
福 建 Fujian	**269**	**23**	**245**	**17**
福 州 Fuzhou	22	3	22	3
厦 门 Xiamen	32	2	28	2
莆 田 Putian				
泉 州 Quanzhou	198	18	178	12
漳 州 Zhangzhou	17		17	
宁 德 Ningde				
山 东 Shandong	**51**	**44**	**42**	**44**
青 岛 Qingdao	8		8	
东 营 Dongying	14	5	11	5
烟 台 Yantai	10	37	6	37
潍 坊 Weifang	6		6	
威 海 Weihai	8		7	
日 照 Rizhao	4	2	3	2
滨 州 Binzhou	1		1	

8-12 续表2 continued

地 区 Region	当年安排施工项目 Arranged for Construction in the Year		当年竣工项目 Completed in the Year	
	治理废水 Treatment of Waste Water	治理固体废物 Treatment of Solid Wastes	治理废水 Treament of Waste Water	治理固体废物 Treatment of Solid Wastes
广 东 Guangdong	**142**	**3**	**136**	**3**
广 州 Guangzhou	6		5	
深 圳 Shenzhen	43	2	43	2
珠 海 Zhuhai	14	1	14	1
汕 头 Shantou	43		41	
江 门 Jiangmen	2		2	
湛 江 Zhanjiang	6		5	
茂 名 Maoming	1			
惠 州 Huizhou	9		8	
汕 尾 Shanwei				
阳 江 Yangjiang	3		3	
东 莞 Dongguan				
中 山 Zhongshan	10		10	
潮 州 Chaozhou	1		1	
揭 阳 Jieyang	4		4	
广 西 Guangxi	**6**		**6**	
北 海 Beihai	3		3	
防城港 Fangchenggang				
钦 州 Qinzhou	3		3	
海 南 Hainan				
海 口 Haikou				
三 亚 Sanya				

8-13 沿海区域海洋类型自然保护区建设情况
Construction of Marine-Type Nature Reserves

地 区 Region	保护区数量（个） Number of Nature Reserves (number)		按保护级别分（个） By Level of Protection (number)		按保护类型分（个） By Type of Protection (number)				保护区面积（平方千米） Area(km²)
	已建 Already Established	新建 Newly Established	国家级 National	地方级 Provincial	海洋和海岸生态系统 Marine and Coastal Ecosystem	海洋自然历史遗迹 Marine Natural and Historical Relics	海洋生物多样性 Marine Biodiversity	其他 Other	
合 计 Total	**125**		**28**	**97**	**30**	**5**	**53**	**37**	**158 400**
环渤海经济区 Round-the-Bohai Sea Economic Zone	29		11	18	1	2	11	15	15 418
长江三角洲经济区 Yangtze River Delta Economic Zone	12		4	8	1	1	7	3	3 130
海峡西岸经济区 Economic Zone on West Side of the Straits	12		3	9	5	2	5		197
珠江三角洲经济区 Zhujiang River Delta Economic Zone	50		4	46	21		29		5 491
环北部湾经济区 Round-the-Beibu Gulf Economic Zone	22		6	16	2		1	19	134 164

8-14 沿海地区海洋类型自然保护区建设情况
Construction of Marine-Type Nature Reserves

地 区 Region	保护区数量（个） Number of Nature Reserves (number)		按保护级别（个） By Level of Protection (number)		按保护类型分（个） By Type of Protection (number)				保护区面积（平方千米） Area(km^2)
	已建 Already Established	新建 Newly Established	国家级 National	地方级 Provincial	海洋和海岸生态系统 Marine and Coastal Ecosystem	海洋自然历史遗迹 Marine Natural and Historical Relics	海洋生物多样性 Marine Biodiversity	其他 Other	
合 计 Total	**125**		**28**	**97**	**30**	**5**	**53**	**37**	**158 400**
天 津 Tianjin	1		1			1			359
河 北 Hebei	3		1	2	1	1	1		339
辽 宁 Liaoning	11		6	5				11	9 370
上 海 Shanghai	4		2	2		1		3	941
江 苏 Jiangsu	5		1	4	1		4		833
浙 江 Zhejiang	3		1	2			3		1 356
福 建 Fujian	12		3	9	5	2	5		197
山 东 Shandong	14		3	11			10	4	5 350
广 东 Guangdong	50		4	46	21		29		5 491
广 西 Guangxi	2		2		2				110
海 南 Hainan	20		4	16			1	19	134 054

8-15 全国海洋生态监控区基本情况
Basic Condition of the Marine Ecological Monitoring Areas Throughout the Country

生态监控区 Ecological Monitoring Area	所在地 Location	面积 （平方千米） Area (km^2)	主要生态系统类型 Major Types of Ecosystem	健康状况 Health Condition
双台子河口	辽宁省	3 000	河 口	亚健康
Shuangtaizi Estuary	Liaoning Province		Estuary	Subhealthy
锦州湾	辽宁省	650	海 湾	不健康
Jinzhou Bay	Liaoning Province		Bay	unhealthy
滦河口-北戴河	河北省	900	河 口	亚健康
Luanhe Mouth-Beidaihe	Hebei Province		Estuary	Subhealthy
渤海湾	天津市	3 000	海 湾	亚健康
Bohai Bay	Tianjin Municipality		Bay	Subhealthy
莱州湾	山东省	3 770	海 湾	亚健康
Laizhou Bay	Shandong Province		Bay	Subhealthy
黄河口	山东省	2 600	河 口	亚健康
Yellow River Mouth	Shandong Province		Estuary	Subhealthy
苏北浅滩	江苏省	15 400	滩涂湿地	亚健康
North Jiangsu Bank	Jiangsu Province		mudflat and Wetland	Subhealthy
长江口	上海市	13 668	河 口	亚健康
Yangtze River Mouth	Shanghai Municipality		Estuary	Subhealthy
杭州湾	上海市 浙江省	5 000	海 湾	不健康
Hangzhou Bay	Shanghai Municipality Zhejiang Province		Bay	unhealthy
乐清湾	浙江省	464	海 湾	亚健康
Yueqing Bay	Zhejiang Province		Bay	Subhealthy

8-15 续表 continued

生态监控区 Ecological Monitoring Area	所在地 Location	面积（平方千米） Area (km^2)	主要生态系统类型 Major Types of Ecosystem	健康状况 Health Condition
闽东沿岸	福建省	5 063	海 湾	亚健康
Constal East Fujian	Fujian Province		Bay	Subhealthy
大亚湾	广东省	1 200	海 湾	亚健康
Daya Bay	Guangdong Province		Bay	Subhealthy
珠江口	广东省	3 980	河 口	亚健康
Zhujiang River Mouth	Guangdong Province		Estuary	Subhealthy
雷州半岛西南沿岸	广东省	1 150	珊瑚礁	亚健康
Southwest Coast of Leizhou Peninsula	Guangdong Province		Coral Reef	Subhealthy
广西北海	广西壮族自治区	120	珊瑚礁 红树林 海草床	健 康 健 康 亚健康
Beihai,Guangxi	Guangxi Zhuang Nationality Autonomous Region		Coral Reef Mangroves Seagrass Bed	Healthy Healthy Subhealthy
北仑河口	广西壮族自治区	150	红树林	亚健康
Beilun River Mouth	Guangxi Zhuang Nationality Autonomous Region		Mangroves	Subhealthy
海南东海岸	海南省	3 750	珊瑚礁 海草床	亚健康 健 康
East Coast of Hainan	Hainan Province		Coral Reef Seagrass Bed	Subhealthy Healthy
西沙珊瑚礁	海南省	400	珊瑚礁	亚健康
Xisha Coral Reef	Hainan Province		Coral Reef	Subhealthy

8-16　沿海地区风暴潮灾害情况
Survey of Storm Surges by Coastal Regions

受灾地区 Disaster Area	受灾人口（万人） Disaster-stricken Population (10 000 persons)	农作物受灾（千公顷） Disaster-Affected crops (1 000 hm^2)	损失船只（艘） Lost Boat (Number)	海水养殖受灾面积（千公顷） Affected Area of Mariculture (1 000 hm^2)
合　计 Total	**437.07**	**38.41**	**743.00**	**57.84**
浙　江 Zhejiang				
福　建 Fujian	116.42	0.00	729	14.79
山　东 Shandong	0.11		14	3.40
广　东 Guangdong	236.30			28.81
广　西 Guangxi	84.24	38.41		0.84

8-16 续表 continued

受灾地区 Disaster Area	损毁海岸工程（千米） Destroyed Coastal Engineering Architectures(km)	死亡人数*（人） Death Toll (person)	直接经济损失（亿元） Direct Economic Loss (100 million yuan)
合　计 Total	**122.49**	**5**	**65.79**
浙　江 Zhejiang	0.50	0	0.01
福　建 Fujian	5.67	0	33.10
山　东 Shandong	2.03	0	0.53
广　东 Guangdong	106.37	5	30.62
广　西 Guangxi	7.92	0	1.53

注：*包括失踪人数。

Notes:*Includes lost people.

8-17 沿海地区赤潮灾害情况
Survey of Red Tide by Coastal Regions

时间 Date	影响区域 Area		最大面积（平方千米）Disaster Area(km^2)
合　计 Total			**9 975**
其中：Including:			
4月15日 Apr.15	江苏省连云港海州湾海域	Haizhou Bay sea area of Lianyungang, Jiangsu Province	120
5月1日至5月10日 May.1-May.10	福建省长乐至平潭沿岸海域	Littoral area from Changle to Pingtan of Fujian Province	620
5月2日 May.2	广西北部湾海域	Beibu Gulf sea area of Guangxi	150
5月4日至5月21日 May.4-May.21	浙江省苍南大渔湾附近海域	Sea area adjacent to Dayu Bay, Cangnan, Zhejiang Province	400
5月4日至5月26日 May.4-May.26	福建省福宁湾、大嵛山、四礵列岛、西洋岛海域	Sea areas of Funing Bay, Dayushan, Sishuang Archipelago and Xiyang Island of Fujian Province	925
5月5日至5月13日 May.5-May.13	福建省莆田南日岛海域	Nanridao sea area, Putian, Fujian Province	600
5月11日至6月1日 May.11-Jun.1	浙江省玉环东部海域	Eastern sea area of Yuhuan, Zhejiang Province	110
5月14日至5月27日 May.14-May.27	浙江省舟山朱家尖东部海域	Eastern sea area of Zhujiajian, Zhoushan, Zhejiang Province	1 040
5月20日至5月29日 May.20-May.29	浙江省南麂附近海域	Waters near Nanji, Zhejiang Province	100
5月26日至6月10日 May.26-Jun.10	浙江省苍南海域	Cannan waters of Zhejiang Province	100
5月30日至6月1日 May.30-Jun.1	浙江省舟山以南海域	Sea area south of Zhoushan, Zhejiang Province	880

8-17 续表 continued

时间 Date	影响区域 Area		最大面积（平方千米） Disaster Area(km^2)
5月30日至6月7日 May.30-Jun.7	江苏省启东以东海域	Sea area to the east of Qidong, Jiangsu Province	400
6月1日至6月9日 Jun.1-Jun.9	福建省福鼎牛栏岗海水浴场至霞浦大京海水浴场	From Niulangang Bathing Beach of Fuding to Dajing Bathing Beach of Xiapu, Fujian Province	180
6月3日至6月12日 Jun.3-Jun.12	天津港锚地附近海域	Sea area near the anchorage of Tianjin Port	140
6月7日至6月13日 Jun.7-Jun.13	福建省连江同心湾、大建湾、黄岐湾海域	Sea areas of Tongxin, Dajian are Huangqi Bays, Lianjiang Fujian Province	150
6月8日至6月11日 Jun.8-Jun.11	浙江舟山泗礁岛东北部海域	Northeast sea area of Sijiao Island, Zhoushan, Zhejiang Province	300
6月24日至7月12日 Jun.24-Jul.12	辽宁省绥中至河北省秦皇岛沿岸海域	Littoral area from Suizhong of Liaoning Province to Qinhuangdao of Hebei Province	3 350
7月5日至7月7日 Jun.24-Jul.12	江苏省连云港海州湾海域	Haizhou Bay sea area of Lianyungang, Jiangsu Province	100
7月21日至7月25日 Jun.21-Jul.25	浙江省象山港海域	Xiangshan Port sea area of Zhejing Province	120
8月4日至8月9日 Aug.4-Aug.9	浙江省象山港海域	Xiangshan Port sea area of Zhejing Province	190

注：本表数据为我国近海面积100平方千米以上的赤潮。

Note: The data is for the red tide over 100 km^2 in China coastal area.

主要统计指标解释

1. 工业废水排放量 指经过企业厂区所有排放口排到企业外部的工业废水量。包括生产废水、外排的直接冷却水、超标排放的矿井地下水和与工业废水混排的厂区生活污水,不包括外排的间接冷却水(清污不分流的间接冷却水应计算在内)。

2. 直接排入海的工业废水量 指经企业位于海边的排放口，直接排入海中的废水量。直接排入是指废水经过工厂的排污口直接排入海，而未经过城市下水道或其他中间体，也不受其他水体的影响。

3. 工业废水处理量 指报告期内各种水治理设施实际处理的工业废水量,包括处理后外排的和处理后回用的工业废水量，虽经处理但未达到国家或地方排放标准的废水量也应计算在内。计算时，如遇车间和厂排放口均有治理设施，并对同一废水分级处理时，不应重复计算工业废水处理量。

4. 工业废水处理率 指工业废水处理量占需要处理的工业废水量的百分率。其计算公式是:

工业废水处理率 =（工业废水处理量 / 需处理的工业废水量）×100%

式中，需处理工业废水量 =工业废水排放量+工业废水处理回用量-(工业废水排放达标量-工业废水处理排放达标量)

5. 工业废水排放达标量 指各项指标都达到国家或地方排放标准的外排工业废水量,包括经过处理后外排达标的和未经处理外排达标的两部分。国家排放标准见 GB 8978-88。

6. 工业废水处理排放达标量 指经过各种水治理设施处理后达到国家或地方排放标准的废水排放量。

7. 工业固体废物处置量 指将固体废物焚烧或者最终置于符合环境保护规定要求的场所并不再回取的工业固体废物量(包括当年处置往年的工业固体废物累计贮存量)。

8. 工业固体废物排放量 指将所产生的固体废物排到固体废物污染防治设施、场所以外的量。不包括矿山开采的剥离废石和掘进废石(煤矸石和呈酸性或碱性的废石除外)。

9. 当年开工污染治理项目数 指报告期内由国家、部门、地方或企业单位安排开工的，并以治理废水、废气、固体废物、噪声和其他(如电磁波、恶臭等)环境污染的环境治理工程的总数。不包括“三同时”项目。

10. 当年竣工项目数 指报告期内竣工投入运行的治理废水、废气、固体废物、噪声及其他污染的环境工程项目的总数。

Explanatory Notes on Main Statistical Indicators

1. Volume of Industrial Waste Water Discharged refers to the quantity of industrial waste water discharged externally through all the outlets in the factory area of the enterprise, including the waste water from production, externally discharged direct cooling water, mine-shaft groundwater discharged exceeding the set standard and the domestic sewage of the factory area discharged together with the industrial waste water, but not including the indirect cooling water discharge externally.

2. Volume of Industrial Waste Water discharged Directly to Sea refers to the quantity of waste water directly discharged into the sea through the outlets of the enterprise by the sea. Direct discharge means the direct discharge into the sea of waste water through the outlets of the factory, which is not discharged via the urban sewers or other intermediates and is not affected by other water bodies.

3. Volume of Industrial Waste Water Treated refers to the industrial waste water volume actually treated by various water treatment facilities in the period covered by the report, including the quantity of the industrial waste water discharged and reused after treatment. The amount of the waste water which is not up to the state or local standard of discharge upon treatment should be included. If the workshops and outlets of the factory are provided with treatment facilities and carry out graded treatment of the same waste water, the processed volume of industrial waste water cannot be calculated repeatedly.

4. Rate of Industrial Waste Water Treatment refers to the percentage of the processed volume of industrial waste water in the industrial waste water volume that needs to be treated. The calculation formula is:

Rate of Industrial Waste Water Treatment = (Processed Volume of Industrial Waste Water/Processed Volume of Industrial Waste Water Required) × 100%.

Where: Processed Volume of Industrial Waste Water Required = (Processed Volume of Industrial Waste Water + Reused Volume of Industrial Waste Water After Treatment−Up-to-Standard Volume of Industrial Waste Water for Discharge−Up-to-Standard Volume of Industrial Waste Water for Discharge after Treatment).

5. Volume of Up-to-Standard Industrial Waste Water Discharged refers to the externally discharged volume of industrial waste water with all its indexes up to the state or local standard for discharge, including the volume of industrial waste water up to the standard for discharge after being treated or that without being treated. See GB8978−88 for the state′s discharge standard.

6. Volume of Up-to-Standard Discharged Industrial Waste Water After Treatment refers to the discharged volume of waste water that is up to the state or local standard for discharge upon treatment by various water treatment facilities.

7. Volume of Industrial Solid Wastes Discharged refers to the amount of solid wastes discharged out of the facilities and sites for the solid wastes pollution prevention and control, not including the stripped and tunneled waste ores in the excavation of mines (other than gangues and acid or alkaline waste ores).

8. Volume of Industrial Solid Wastes Treated refers to the volume of industrial solid wastes which are to be burned or finally placed at the sites in keeping with the requirement of environmental protection and will not be recovered (including the accumulated amount of storage in former years of industrial solid radioactive matter disposed of in the year).

9. Number of Pollution Treatment Projects Started in the Current Year refers to the total number of environmental pollution control projects for controlling waste water, waste gas, solid wastes, noise and other environmental pollutions (such as electromagnetic wave, offensive odor) started by the state, governmental departments, local governments or enterprises in the period covered by the report mainly for the purpose of pollution control and multipurpose utilization of ″three wastes″.

10. Number of Pollution Treatment Projects Completed in the Current Year refers to the total number of environmental engineering projects for controlling waste water, waste gas, solid wastes, noise and other pollutions which are completed and put into operation in the period covered by the report.

9 海洋行政管理及公益服务

Marine Administration and Public-Good Service

9-1 海域使用管理情况
Sea Area Use Management

地　区 Region	发放海域使用权证书（本） Certificates of Right of Sea Area Use Issued (number)	确权海域面积（公顷） Area of Waters with Established Rights (hm^2)	海域使用金（万元） Charge for Sea Area Utilization (10 000 yuan)
全国总计 National Total	**2 481**	**193 769.16**	**907 374.33**
天 津 Tianjin	51	2 150.02	200 012.32
河 北 Hebei	52	4 679.83	51 951.70
辽 宁 Liaoning	632	105 470.49	166 901.71
上 海 Shanghai	4	113.21	882.14
江 苏 Jiangsu	206	35 657.93	51 063.93
浙 江 Zhejiang	218	4 872.82	90 786.21
福 建 Fujian	253	6 537.53	91 439.51
山 东 Shandong	395	22 445.01	108 425.10
广 东 Guangdong	352	6 762.70	70 409.34
广 西 Guangxi	208	2 163.12	50 627.53
海 南 Hainan	94	1 338.70	15 021.66
其 他* other	16	1 577.80	9 853.18

注：*为沿海省（自治区、直辖市）管理海域以外。

Note: * Sea areas outside the control of coastal provinces of autonomous resgions and municipalities directly under the Central Government.

9-2 海洋倾废管理情况
Management on the Ocean Dumping of Wastes

海　区 Sea Area	签发疏浚物海洋倾倒许可证（份） Permit Issued for Ocean Dumping of Dredged Materials (number)	新选划倾倒区数（个） Number of Newly Designated Dumping Zones (number)
合　计 Total	**322**	**45**
渤黄海 Bohai and Huangai Sea	34	22
东　海 East China Sea	240	12
南　海 South China Sea	48	11

9-3 海洋执法检查情况
Marine Law Enforcement Inspection

执法检查项目分类 Classification of Items Subject to Low Enforcement Inspection	检查项目（个） Number of Items Subject to Inspection (number)	检查次数（次） Number of Times of Inspection (number)	发现违法行为（起） Illegal Acts Found (case)
合　计 Total	**29 176**	**72 233**	**1 836**
渔业用海 Sea Use for Fishery	19 725	39 949	1 023
工业用海 Sea Use for Industrial Purposes	2 437	6 022	232
交通运输用海 Sea Use for Transportation and Communications	2 293	7 173	98
旅游娱乐用海 Sea Use for Tourism and Recreation	654	5 626	40
海底工程用海 Sea Use for Undersea Engineering	140	793	0
排污倾倒用海 Sea Use for Sewage Discharge and Dumping	510	1 668	36
造地工程用海 Sea Use for Reclaiming Land from Sea	2 731	8 804	278
特殊用海 Special Sea Uses	151	666	22
其他用海 Other Sea Uses	535	1 532	107

9-4 沿海地区海滨观测台站分布概况
Distribution of Coastal Observation Stations by Coastal Regions

单位：个 (number)

地 区 Region	合 计 Total	海洋站 Marine Station	验潮站① Tide Station	气象台站② Meteorological Station	地震台站 Seismic Station	雷达站 Radar Station
合 计 Total	**975**	**105**	**334**	**110**	**284**	**143**
天 津 Tianjin	**15**	1	1	3	6	4
河 北 Hebei	**42**	4	4	7	21	6
辽 宁 Liaoning	**68**	7	10	11	29	11
上 海 Shanghai	**95**	6	65	0	10	14
江 苏 Jiangsu	**56**	4	23	6	19	4
浙 江 Zhejiang	**131**	21	51	18	22	19
福 建 Fujian	**120**	15	24	15	45	21
山 东 Shandong	**133**	17	36	16	41	23
广 东 Guangdong	**221**	16	98	27	54	26
广 西 Guangxi	**33**	5	10	1	11	6
海 南 Hainan	**61**	9	12	5	26	9

注： ①潮流量观测站34处，潮水位观测站300处。
②中国气象局所属的气象台站仅指沿海气象台站中有海洋气象观测、预报及服务等业务的台站，合计中含有中央气象台1个。

Notes: ①There are 34 tidal current observation stations and 300 tidal level observation stations.
② The meteorological observatories and stations under the China Meteorological Administration only refer to those with marine meteorotogical observation, forecast and service among the coastal meteorological observatories and the total includes the central meteorological station.

9-5 海洋预报服务概况
Marine Forecast Service

单位：次 (time)

预报项目 Item	数值预报 Numerical Forecast			
	预报服务次数 Frequency	发布次数 Frequency of Release		
		广播电视 Radio and TV	因特网 Internet	纸 质 Paper Media
合　计 Total	**17 683**	**528**	**17 142**	**1 043**
海　浪 Sea Wave	2 865		3 577	18
海　温 Sea Surface Temperature	2 242	52	2 190	52
潮　汐 Tide	3 130		3 130	
海　流 Sea Current	3 022	52	2 970	
海平面 Sea Level				
盐　度 Salinity	730		730	
赤　潮 Red Tide				
海水浴场 Bathing Beach				
海　冰 Sea Ice	420		84	
风暴潮 Storm Surge	2 614		2 467	17
气　象 Metorology	1 752		1 752	
厄尔尼诺 EL Niño	12		12	720
专　项 Special Item	13		3	10
其　他 Others	883	424	227	226

9-5 续表 continued

预报项目 Item	统计预报 Statistical Forecast			
	预报服务次数 Frequency	发布次数 Frequency of Release		
		广播电视 Radio and TV	因特网 Internet	纸 质 Paper Media
合 计 Total	**154 103**	**55 006**	**69 940**	**81 485**
海 浪 Sea Wave	40 691	16 837	20 143	16 422
海 温 Sea Surface Temperature	31 802	15 971	16 757	11 718
潮 汐 Tide	30 865	15 312	14 612	10 507
海 流 Sea Current	1 700	365	850	485
海平面 Sea Level				
盐 度 Salinity	34			34
赤 潮 Red Tide	238	43	36	162
海水浴场 Bathing Beach	2 786	2 795	1 572	2 253
海 冰 Sea Ice	5 199	519	547	1 043
风暴潮 Storm Surge	2 798	169	714	7 076
气 象 Metorology	10 652	72	4 885	8 935
厄尔尼诺 EL Niño	13		13	721
专 项 Special Item	23 475	1 171	8 048	18 394
其 他 Others	3 850	1 752	1 763	3 735

注：本表只包含国家海洋局资料。

Note: This table only contains the data from the State Oceanic Administration.

9-6 海洋环境观测情况
Condition of Marine Environmental Observation

项目 Item	台站观测 Station Observation	断面观测 Sectional Observation	浮标观测 Buoy Observation	船舶测报 Ship Measuring and Reporting	其他观测 Other Observation
测站（点）数（个） Number of Station (unit)	105	121	18	130	183
实际获得数据量（个） Quantity of Data Actually Obtained (unit)	93 093 678 *	17 689	1 456 244	7 519 487	140 024 833

注：*为不完全统计数据。
Note: * is incomplete statistial data.

9-7 海洋调查概况
Marine Survey Statistics

调查名称 Name	站点数（个） Number of Stations (unit)	船舶数（艘） Number of Ships (unit)	项目数（个） Number of Items (unit)	实际获得数据（个） Quantity of Data Actually Obtained (unit)	发布通（公、简）报量（期） Quantity of Circulars (Bulletin, Brief Reports) (number)	提交报告数（份） Number of Reports Submitted (unit)
合 计 Total	**1 517**	**65**	**4 230**	**41 640 371**	**103**	**98**
大洋调查 Oceanic Survey	388	1	80	40 000 000	20	21
极地调查 Polar Survey	168	5	29	942		5
专项调查 Special Survey	118	11	67	1 603 095	8	33
其他调查 Other Survey	843	48	4 054	36 334	75	39

9-8 涉外海洋科学研究审批情况
Examination and Approval on Foreign-Related Marine Scientific Research

审批单位 Examing and Approving Units	审批研究活动申请（份） Application for the Examing and Approving Research Activity (unit)	船只作业计划审批（份） Examination and Approval of the Ship Operating Plan (unit)
国家海洋局 State Oceanic Administration , People's Republic of China	8	8

9-9 海洋档案及利用情况
Marine File and Its Use

指　标 Item	指 标 值 Data
室（馆）存档案 Files Deposited in the Archives	
纸介质（卷、册） Paper Media (reel,volume)	118 334
磁介质（盘） Magnetic Media (spool)	8 992
科技档案利用 Utilization of Scientific as Technological Files	
接待读者（人次） Readers Received (person-time)	2 814
借阅案卷、盘 Borrowed Files and Spools	
纸介质（卷次） Paper Media (Number of Reels)	3 622
电子（盘次） Electronic Document (spool-times)	1 630

9-10 卫星遥感接收应用情况
Remote-Sensing Receiving and Utilization

指　标 Item	指标值 Data
卫星接收次数（次） Number of Satellite Receptions (time)	136 197
全年接收时间（分钟） Whole Year Receiving Time (minute)	5 340 904
全年实际接收存档数据量（GB） Amount of Data Actually Received and Placed on File in the Whole Year (GB)	28 772.96
累计存档数据量（GB） Total Amount of Data Placed on File (GB)	38 843.50
卫星数据分发 Satellite Data Distribution	
类别用户（个） 　　Classified Users (number)	216
分发数据量（GB） 　　Amount of Data Distributed (GB)	15 023.79

9-11 海洋标准化监督管理情况
Supervision and Management of Marine Standardization

单位：项 (item)

指 标 Item	指标值 Data
标准立项审查 Examination of the Standards for Authorization	
国家标准 National Standards	33
行业标准 Professional Standards	257
标准审查 Examination of Standards	
国家标准 National Standards	78
行业标准 Professional Standards	155
标准出版 Standards Publication	
国家标准 National Standards	2
行业标准 Professional Standards	13
标准实施监督检查（次） Implementation Supervision and Examination of Standards (time)	6

主要统计指标解释

1. 海域使用检查　针对不同类型的用海行为进行的监督检查。

2. 涉外海洋科研项目检查　主要针对国际组织、外国组织和个人为和平目的，单独或者与中华人民共和国的组织合作，使用船舶或者其他运载工具、设施，在中华人民共和国内海、领海以及中华人民共和国管辖的其他海域内进行的对海洋环境和海洋资源等的调查研究活动进行监督检查。

3. 海底电缆管道检查　主要是针对铺设海底电缆管道路由调查、铺设施工和维修改造等的监督检查。

4. 海洋工程建设项目环境保护检查　主要是针对防治海洋工程建设项目对海洋环境的污染损害的监督检查。

5. 海洋倾废检查　主要是针对防治倾倒废弃物对海洋环境的污染损害的监督检查。

6. 海洋生态保护检查　主要是针对红树林、珊瑚礁、滨海湿地、海岛、海湾、入海河口、重要渔业水域等具有典型性、代表性的海洋生态系统，珍稀、濒危海洋生物的天然集中分布区，具有重要经济价值的海洋生物生存区域及有重大科学文化价值的海洋自然历史遗迹和自然景观等海洋自然保护区以及其他需要予以特殊保护的区域的监督检查。

7. 断面监测　按照国家海洋局“断面监测方案”，每年定期利用船舶在沿海设定的断面上进行海洋水文、气象、生物、化学等项目的监测活动。

8. 浮标监测　在海上固定站位获取长期、连续海洋环境观测资料的海上锚定资料浮标。

9. 船舶监测　利用在固定航线上走航的商船或渔船，每日定时所在地的海洋环境状况（主要是水文气象要素），并将监测数据实时发送给有关单位，供海洋环境预报使用。

10. 大洋调查　以大洋科考、研究为目的的远洋调查。

11. 专项调查　为完成国家专项任务进行的海洋调查。

12. 全年接收时间　指全年整个应用系统接收的各类遥感卫星数据时，接收设备工作时间。

13. 全年实际接收存档数据量　指全年整个应用系统接收的各类遥感卫星数据原始数据量。

14. 累计存档数据量　指全年整个应用系统制作的各类遥感产品数据量。

15. 分发数据量　指应用系统提供数据产品用于开展各类应用的数据量。

16. 卫星接收次数　地面观测系统接收卫星数据的数量。

17. 国家标准　针对海洋领域内需要在全国范围内统一的有关技术要求所制定的国家标准。海洋国家标准由国家标准化主管部门统一批准、编号和发布。

18. 行业标准　对没有海洋国家标准而又需要在海洋领域内统一的技术要求所制定的标准。海洋行业标准由国家海洋局统一批准、编号和发布。

19. 获奖成果数　指本年度内机构作为第一完成单位从地（市）以上政府科技管理部门获得的各种科技成果奖的项数。

20. 国家级奖励　指国家自然科学奖、国家发明奖、国家科技进步奖。

21. 省部级奖励　指以国务院各部门名义颁发的或以省、自治区、直辖市政府（科委）名义颁发的重大科技成果奖和科技进步奖等。
22. 地市级奖励　指以地（市、州）政府（科委）和省属各部门名义颁发的重大科技成果奖和科技进步奖等。

Explanatory Notes on Main Statistical Indicators

1. Sea Area Use Supervision and inspection refers to the various types of sea area use conducts.
2. Inspection of Foreign-Related Marine Scientific Research Projects refers to the supervision and inspection of the activities of surveying marine environment and resources conducted by international organizations, foreign organizations and individuals for peaceful purposes, alone or in cooperation with PRC organizations, by using ships or other means of delivery as facilities in the internal seas and territorial waters of the People's Republic of China as well as in the other water under the jurisdiction of the People's Republic of China.
3. Inspection of Submarine Cables and Pipelines is mainly aimed at the survey of the submarine cables and pipelines laying route and supervision and inspection of the laying construction, repair and transformation.
4. Environmental Protection Inspection for the Marine Engineering Construction Projects is mainly directed against preventing and controlling the pollution damage of the marine engineering construction project to the marine environment.
5. Inspection of Oceanic Dumping of Wastes mainly refers to the supervision and inspection aimed at preventing and controlling the pollution damage of wastes dumping to the marine environment.
6. Inspection of Marine Ecological Protection mainly refers to the supervision and inspection of the typical and representative marine ecosystems such as mangroves, coral reef, coastal wetland, sea island, bay, estuaries open to the sea, important fishery waters, etc, the natural centralized distribution zones of rare and endangered marine life, the living areas for the marine life with significant economic values and the marine nature reserves such as the marine natural and historical remains and natural landscapes with important scientific and cultural values as well as other areas needing to be specially protected.
7. Sectional Monitoring According to the "*Sectional Monitoring Plan*" of the State Oceanic Administration, monitoring activities concerning such items as marine hydrology, meteorology, biology and chemistry are carried out regularly every year on the sections set in the coastal area.
8. Buoy Monitoring refers to the monitoring carried out by the offshore mooring data buoys which acquire long-term, continuous marine environmental observations at the fixed stations at sea.
9. Ship Monitoring Monitoring the marine environmental condition in the sea area of fixed times every day by using the merchant or fishing vessels cruising on the fixed navigation line (mainly the

hydro meteorological elements) and transmitting the monitored data in real time to the related units for use in the marine environmental forecast.

10. Oceanic Survey refers to the oceanic surveys aimed at the oceanic scientific investigations and research.

11. Special-Subject Survey refers to the oceanic investigation for the purpose of fulfilling the state′s special tasks

12. Whole Year Receiving Time refers to the operating time of receiving equipment in the whole year when the whole application system receives all types of remote sensing satellite data.

13. Amount of Data Actually Received and Placed on File Throughout the year refers to the raw data of remote-sensing satellite data of various kinds received by the whole application system throughout the year.

14. Total Amount of Date Placed on File refers to the date amount of various remote-sensing products made by the whole application systems throughout the year.

15. Amount of Data Distributed refers to the amount of data products provided by the application system for various uses.

16. Number of Satellite Receptions refers to the amount of data received by the ground observation system.

17. National Standards refers to the standards formulated in view of the relevant technical requirements in the marine field that need to be unified throughout the country. The Marine National standards are approved, numbered and issued uniformly by the state department responsible for standardization.

18. Professional Standards refers to the standards formulated for the technical requirements which have no national standards but need to be unified in the marine field. The Marine Professional Standards are approved, numbered and issued by the State Oceanic Administration.

19. Number of Prize-Winning Results refers to the number of items of various scientific and technological results awards obtained from the departments of scientific and technological management of the governments above the prefecture (city) level by the institution and the first accomplishing unit.

20. State-Level Awards refers to the State Natural Science Award, State Invention Prize and State Scientific and Technological Progress Prize.

21. Province(Ministry)-Level Awards refers to the significant scientific and technological achievement prize and the scientific and technological progress prize awarded in the name of the departments of the State Council or the governments of provinces, autonomous regions and municipalities directly under the Central Government (commissions or science and technology).

22. Prefecture-Level Awards refers to the significant scientific and technological achievements prize and the scientific and technological progress prize awarded in the name of the prefecture government (commission of science and technology) and the various departments of the province.

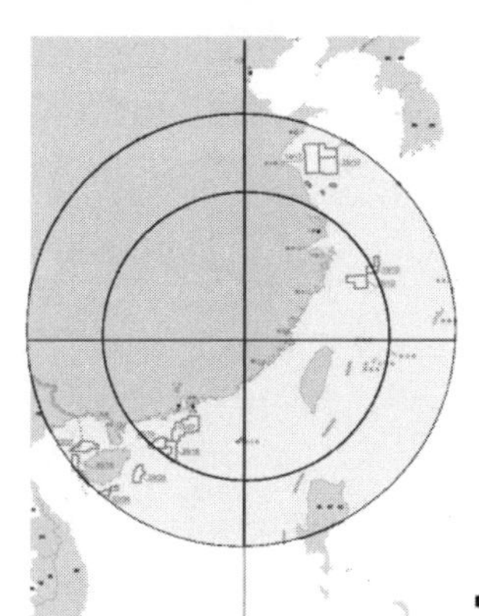

10

全国及沿海社会经济

National and Coastal Socioeconomy

10-1 国内生产总值
Gross Domestic Product

单位：亿元　　　　(100 million yuan)

年　份 Year	国内生产总值 Gross Domestic Product	第一产业 Primary Industry	第二产业 Secondary Industry	第三产业 Tertiary Industry
2001	109 655.2	15 781.3	49 512.3	44 361.6
2002	120 332.7	16 537.0	53 896.8	49 898.9
2003	135 822.8	17 381.7	62 436.3	56 004.7
2004	159 878.3	21 412.7	73 904.3	64 561.3
2005	184 937.4	22 420.0	87 598.1	74 919.3
2006	216 314.4	24 040.0	103 719.5	88 554.9
2007	265 810.3	28 627.0	125 831.4	111 351.9
2008	314 045.4	33 702.0	149 003.4	131 340.0
2009	340 902.8	35 226.0	157 638.8	148 038.0
2010	401 202.0	40 533.6	187 581.4	173 087.0

10-2 国内生产总值增长速度
Growth Rate of Gross Domestic Product

年　份 Year	国内生产总值(%) Gross Domestic Product(%)	第一产业 Primary Industry	第二产业 Secondary Industry	第三产业 Tertiary Industry
2001	8.3	2.8	8.4	10.3
2002	9.1	2.9	9.8	10.4
2003	10.0	2.5	12.7	9.5
2004	10.1	6.3	11.1	10.1
2005	11.3	5.2	12.1	12.2
2006	12.7	5.0	13.4	14.1
2007	14.2	3.7	15.1	16.0
2008	9.6	5.4	9.9	10.4
2009	9.2	4.2	9.9	9.6
2010	10.4	4.3	12.4	9.6

注：本表按可比价格计算(上年为基期)。

Note:This table is calculated at the comparable price (with the previous year as the base period).

10-3 沿海地区生产总值
Gross Regional Product of Coastal Regions

单位：亿元 (100 million yuan)

地 区 Region	地区生产总值 Gross Regional Product	第一产业 Primary Industry	第二产业 Secondary Industry	第三产业 Tertiary Industry
合 计 Total	**245 944.20**	**17 808.10**	**125 653.44**	**102 482.66**
天 津 Tianjin	9 224.46	145.58	4 840.23	4 238.65
河 北 Hebei	20 394.26	2 562.81	10 707.68	7 123.77
辽 宁 Liaoning	18 457.27	1 631.08	9 976.82	6 849.37
上 海 Shanghai	17 165.98	114.15	7 218.32	9 833.51
江 苏 Jiangsu	41 425.48	2 540.10	21 753.93	17 131.45
浙 江 Zhejiang	27 722.31	1 360.56	14 297.93	12 063.82
福 建 Fujian	14 737.12	1 363.67	7 522.83	5 850.62
山 东 Shandong	39 169.91	3 588.28	21 238.49	14 343.14
广 东 Guangdong	46 013.06	2 286.98	23 014.53	20 711.55
广 西 Guangxi	9 569.85	1 675.06	4 511.68	3 383.11
海 南 Hainan	2 064.50	539.83	571.00	953.67

10-4 沿海地区生产总值增长速度
Growth Rate of Gross Regional Product of Coastal Regions

单位：%　　(%)

地　区 Region	2001	2002	2003	2004	2005	2006	2007	2008	2009	2010
天　津 Tianjin	12.0	12.7	14.8	15.8	14.9	14.7	15.5	16.5	16.5	17.4
河　北 Hebei	8.7	9.6	11.6	12.9	13.4	13.4	12.8	10.1	10.0	12.2
辽　宁 Liaoning	9.0	10.2	11.5	12.8	12.7	14.2	15.0	13.4	13.1	14.2
上　海 Shanghai	10.5	11.3	12.3	14.2	11.4	12.7	15.2	9.7	8.2	10.3
江　苏 Jiangsu	10.2	11.7	13.6	14.8	14.5	14.9	14.9	12.7	12.4	12.7
浙　江 Zhejiang	10.6	12.6	14.7	14.5	12.8	13.9	14.7	10.1	8.9	11.9
福　建 Fujian	8.7	10.2	11.5	11.8	11.6	14.8	15.2	13.0	12.3	13.9
山　东 Shandong	10.0	11.7	13.4	15.4	15.0	14.7	14.2	12.0	12.2	12.3
广　东 Guangdong	10.5	12.4	14.8	14.8	14.1	14.8	14.9	10.4	9.7	12.4
广　西 Guangxi	8.3	10.6	10.2	11.8	13.1	13.6	15.1	12.8	13.9	14.2
海　南 Hainan	9.1	9.6	10.6	10.7	10.5	13.2	15.8	10.3	11.7	16.0

注：本表按可比价格计算(上年为基期)。

Note: This table is calculated at the comparable price (with the previous year as the base period).

10-5 沿海城市生产总值（2009年）
Gross Regional Product of Coastal Cities, 2009

单位：亿元 (100 million yuan)

沿海城市 Coastal City		地区生产总值 Gross Regional Product	第一产业 Primary Industry	第二产业 Secondary Industry	第三产业 Tertiary Industry
合　计	**Total**	**122 060.7**	**6 927.5**	**59 957.2**	**55 176.0**
天　津	**Tianjin**	**7 521.9**	**128.9**	**3 987.8**	**3 405.2**
河　北	**Hebei**	**6 418.6**	**678.8**	**3 382.8**	**2 357.0**
唐　山	Tangshan	3 812.7	360.2	2 202.1	1 250.4
秦皇岛	Qinhuangdao	804.6	102.4	311.7	390.5
沧　州	Cangzhou	1 801.3	216.2	869.0	716.1
辽　宁	**Liaoning**	**7 613.6**	**736.6**	**3 817.4**	**3 059.6**
大　连	Dalian	4 349.5	313.4	2 127.3	1 908.8
丹　东	Dandong	607.5	87.0	292.8	227.7
锦　州	Jinzhou	727.3	133.0	335.5	258.8
营　口	Yingkou	806.9	67.5	454.3	285.1
盘　锦	Panjin	676.9	73.8	412.9	190.2
葫芦岛	Huludao	445.5	61.9	194.6	189.0
上　海	**Shanghai**	**15 046.5**	**113.8**	**6 001.8**	**8 930.9**
江　苏	**Jiangsu**	**5 731.0**	**721.4**	**2 966.8**	**2 042.8**
南　通	Nantong	2 872.8	236.5	1 607.5	1 028.8
连云港	Lianyungang	941.2	154.5	435.6	351.1
盐　城	Yancheng	1 917.0	330.4	923.7	662.9
浙　江	**Zhejiang**	**18 813.4**	**870.7**	**9 837.4**	**8 105.3**
杭　州	Hangzhou	5 087.5	190.5	2 387.1	2 509.9
宁　波	Ningbo	4 329.2	183.5	2 362.1	1 783.6
温　州	Wenzhou	2 527.3	80.3	1 314.2	1 132.8
嘉　兴	Jiaxing	1 918.0	107.5	1 112.5	698.0
绍　兴	Shaoxing	2 375.8	124.5	1 361.1	890.2
舟　山	Zhoushan	535.2	52.2	242.7	240.3
台　州	Taizhou	2 040.4	132.2	1 057.7	850.5
福　建	**Fujian**	**9 892.7**	**788.1**	**4 848.4**	**4 256.2**
福　州	Fuzhou	2 604.1	242.0	1 108.2	1 253.9
厦　门	Xiamen	1 737.2	20.5	821.0	895.7
莆　田	Putian	691.4	76.6	375.0	239.8
泉　州	Quanzhou	3 069.5	116.7	1 778.7	1 174.1
漳　州	Zhangzhou	1 178.1	218.7	520.0	439.4
宁　德	Ningde	612.4	113.6	245.5	253.3

10-5 续表 continued

沿海城市 Coastal City		地区生产总值 Gross Regional Product	第一产业 Primary Industry	第二产业 Secondary Industry	第三产业 Tertiary Industry
山 东	**Shandong**	**17 274.8**	**1 252.4**	**9 931.7**	**6 090.7**
青 岛	Qingdao	4 853.9	230.3	2 420.1	2 203.5
东 营	Dongying	2 058.9	74.7	1 521.9	462.3
烟 台	Yantai	3 701.8	285.9	2 227.3	1 188.6
潍 坊	Weifang	2 707.3	302.0	1 526.1	879.2
威 海	Weihai	1 733.2	136.3	1 001.0	595.9
日 照	Rizhao	864.7	87.3	471.9	305.5
滨 州	Binzhou	1 355.0	135.9	763.4	455.7
广 东	**Guangdong**	**32 102.0**	**1 331.2**	**14 644.6**	**16 126.2**
广 州	Guangzhou	9 138.3	172.3	3 405.2	5 560.8
深 圳	Shenzhen	8 201.4	6.7	3 827.1	4 367.6
珠 海	Zhuhai	1 038.7	28.8	544.0	465.9
汕 头	Shantou	1 035.9	56.9	569.7	409.3
江 门	Jiangmen	1 341.0	104.4	777.5	459.1
湛 江	Zhanjiang	1 156.7	249.3	445.9	461.5
茂 名	Maoming	1 231.2	240.4	460.8	530.0
惠 州	Huizhou	1 414.8	90.3	789.0	535.5
汕 尾	Shanwei	390.0	67.6	173.6	148.8
阳 江	Yangjiang	527.3	122.3	215.2	189.8
东 莞	Dongguan	3 763.9	14.8	1 823.1	1 926.0
中 山	Zhongshan	1 566.4	45.3	904.4	616.7
潮 州	Chaozhou	480.3	35.6	263.9	180.8
揭 阳	Jieyang	816.1	96.5	445.2	274.4
广 西	**Guangxi**	**968.3**	**231.0**	**384.7**	**352.6**
北 海	Beihai	321.1	77.1	118.4	125.6
防城港	Fangchenggang	251.0	39.9	124.9	86.2
钦 州	Qinzhou	396.2	114.0	141.4	140.8
海 南	**Hainan**	**677.9**	**74.6**	**153.8**	**449.5**
海 口	Haikou	495.3	40.7	114.4	340.2
三 亚	Sanya	182.6	33.9	39.4	109.3

注：本表各省数据为合计数。

Note: The data for the provinces are the totals.

10-6 沿海县生产总值（2009年）
Gross Regional Product of Coastal Counties, 2009

单位：万元　　　　(10 000 yuan)

沿海县 Coastal County		地区生产总值 Gross Regional Product	第一产业 Primary Industry	第二产业 Secondary Industry	第三产业 Tertiary Industry
合 计	**Total**	**257 776 384**	**34 744 706**	**134 989 735**	**88 041 943**
河 北	**Hebei**	**8 789 463**	**1 964 131**	**3 680 275**	**3 145 057**
滦 南	Luannan	2 267 838	495 891	1 025 709	746 238
乐 亭	Leting	2 182 405	498 119	942 788	741 498
唐 海	Tanghai	639 035	122 674	258 728	257 633
昌 黎	Changli	1 054 336	351 557	392 149	310 630
抚 宁	Funing	1 100 805	306 129	463 778	330 898
黄 骅	Huanghua	1 353 086	145 605	516 598	690 883
海 兴	Haixing	191 958	44 156	80 525	67 277
辽 宁	**Liaoning**	**22 194 641**	**4 311 677**	**11 444 492**	**6 438 472**
长 海	Changhai	420 823	286 500	51 775	82 548
瓦房店	Wafangdian	5 082 008	645 300	2 855 866	1 580 842
普兰店	Pulandian	3 924 636	587 369	2 412 597	924 670
庄 河	Zhuanghe	3 946 682	737 559	2 092 733	1 116 390
东 港	Donggang	2 739 192	420 054	1 317 356	1 001 782
凌 海	Linghai	1 361 379	342 283	703 729	315 367
盖 州	Gaizhou	1 200 276	227 421	549 387	423 468
大 洼	Dawa	1 418 809	408 206	733 615	276 988
盘 山	Panshan	875 009	306 070	371 754	197 185
绥 中	Suizhong	747 889	240 720	202 734	304 435
兴 城	Xingcheng	477 938	110 195	152 946	214 797
江 苏	**Jiangsu**	**29 659 700**	**4 985 300**	**14 726 700**	**9 947 700**
海 安	Hai'an	3 007 500	347 000	1 630 000	1 030 500
如 东	Rudong	2 975 000	407 700	1 570 700	996 600

10-6 续表1 continued

沿海县 Coastal County	地区生产总值 Gross Regional Product	第一产业 Primary Industry	第二产业 Secondary Industry	第三产业 Tertiary Industry
启　东 Qidong	3 595 000	492 900	1 884 800	1 217 300
海　门 Haimen	4 150 000	334 300	2 479 700	1 336 000
赣　榆 Ganyu	1 824 400	328 400	842 200	653 800
东　海 Donghai	1 626 900	361 400	695 700	569 800
灌　云 Guanyun	1 193 600	338 400	529 600	325 600
灌　南 Guannan	1 071 000	243 600	502 800	324 600
响　水 Xiangshui	1 045 700	242 900	509 100	293 700
滨　海 Binhai	1 596 000	353 100	690 900	552 000
射　阳 Sheyang	2 010 700	487 000	825 400	698 300
东　台 Dongtai	3 136 300	574 200	1 467 600	1 094 500
大　丰 Dafeng	2 427 600	474 400	1 098 200	855 000
浙　江 Zhejiang	**61 693 621**	**4 096 706**	**35 019 383**	**22 577 532**
象　山 Xiangshan	2 361 096	367 910	1 140 146	853 040
宁　海 Ninghai	2 362 031	252 525	1 321 200	788 306
余　姚 Yuyao	4 891 914	295 319	2 889 840	1 706 755
慈　溪 Cixi	6 249 362	313 719	3 711 013	2 224 630
奉　化 Fenghua	1 938 583	187 824	934 402	816 357
洞　头 Dongtou	300 403	33 975	113 223	153 205
平　阳 Pingyang	1 738 694	93 056	840 772	804 866
苍　南 Cangnan	2 172 262	171 506	1 030 529	970 227
瑞　安 Rui'an	3 828 387	124 394	1 927 390	1 776 603
乐　清 Yueqing	4 235 165	146 210	2 559 541	1 529 414
海　盐 Haiyan	2 102 317	151 909	1 363 951	586 457
海　宁 Haining	3 752 616	178 143	2 268 766	1 305 707
平　湖 Pinghu	2 837 232	138 738	1 805 482	893 012
绍　兴 Shaoxing	6 557 783	240 505	4 056 179	2 261 099
上　虞 Shangyu	3 681 522	261 784	2 169 438	1 250 300
岱　山 Daishan	1 023 587	149 214	580 576	293 797

10-6 续表2 continued

沿海县 Coastal County		地区生产总值 Gross Regional Product	第一产业 Primary Industry	第二产业 Secondary Industry	第三产业 Tertiary Industry
嵊 泗	Shengsi	534 218	93 234	233 200	207 784
玉 环	Yuhuan	2 447 637	165 812	1 503 720	778 105
三 门	Sanmen	888 824	134 107	399 790	354 927
温 岭	Wenling	5 020 099	360 766	2 692 411	1 966 922
临 海	Linhai	2 769 889	236 056	1 477 814	1 056 019
福 建	**Fujian**	**39 266 400**	**4 458 100**	**20 907 800**	**13 900 500**
连 江	Lianjiang	1 577 500	572 300	504 700	500 500
罗 源	Luoyuan	788 800	162 800	460 800	165 200
平 潭	Pingtan	735 700	226 900	134 600	374 200
福 清	Fuqing	4 113 600	559 200	2 004 200	1 550 200
长 乐	Changle	2 546 700	243 900	1 588 800	714 000
仙 游	Xianyou	1 193 800	183 000	521 400	489 400
惠 安	Hui'an	3 445 200	208 000	2 017 000	1 220 200
石 狮	Shishi	3 253 200	127 100	1 766 600	1 359 500
晋 江	Jinjiang	7 988 900	132 200	5 108 700	2 748 000
南 安	Nan'an	4 134 300	167 400	2 563 600	1 403 300
云 霄	Yunxiao	583 200	182 900	169 400	230 900
漳 浦	Zhangpu	1 256 000	373 100	380 700	502 200
诏 安	Zhao'an	809 100	255 600	281 700	271 800
东 山	Dongshan	589 900	171 500	231 000	187 400
龙 海	Longhai	2 947 900	338 100	1 709 600	900 200
霞 浦	Xiapu	798 700	200 700	221 100	376 900
福 安	Fu'an	1 428 300	198 300	739 200	490 800
福 鼎	Fuding	1 075 600	155 100	504 700	415 800
山 东	**Shandong**	**67 317 641**	**6 223 040**	**38 718 700**	**22 375 901**
胶 州	Jiaozhou	5 151 024	329 124	2 924 700	1 897 200
即 墨	Jimo	5 360 770	400 970	2 915 700	2 044 100

10-6 续表3 continued

沿海县 Coastal County		地区生产总值 Gross Regional Product	第一产业 Primary Industry	第二产业 Secondary Industry	第三产业 Tertiary Industry
胶　南	Jiaonan	4 833 200	366 300	2 815 200	1 651 700
垦　利	Kenli	1 872 790	118 903	1 364 562	389 325
利　津	Lijin	1 173 415	177 420	705 465	290 530
广　饶	Guangrao	3 773 243	264 544	2 795 893	712 806
长　岛	Changdao	410 280	240 005	41 281	128 994
龙　口	Longkou	6 380 054	292 270	4 042 488	2 045 296
莱　阳	Laiyang	3 075 362	284 568	1 832 473	958 321
莱　州	Laizhou	4 554 077	417 932	2 730 003	1 406 142
蓬　莱	Penglai	3 231 461	209 185	2 038 184	984 092
招　远	Zhaoyuan	1 783 181	220 479	263 834	1 298 868
海　阳	Haiyang	2 044 633	400 179	964 237	680 217
寿　光	Shouguang	4 167 364	606 393	2 046 970	1 514 001
昌　邑	Changyi	2 012 317	254 637	1 176 783	580 897
文　登	Wendeng	5 648 462	370 941	3 233 916	2 043 605
荣　成	Rongcheng	6 134 568	541 710	3 603 893	1 988 965
乳　山	Rushan	3 212 650	245 627	1 920 980	1 046 043
无　棣	Wudi	1 548 817	256 164	892 214	400 439
沾　化	Zhanhua	949 973	225 689	409 924	314 360
广　东	**Guangdong**	**20 374 364**	**4 998 397**	**8 363 878**	**7 012 089**
南　澳	Nan'ao	80 190	26 289	29 905	23 996
台　山	Taishan	1 929 543	246 290	1 122 807	560 446
恩　平	Enping	814 677	115 132	274 464	425 081
遂　溪	Suixi	1 165 085	530 032	278 237	356 816
徐　闻	Xuwen	654 005	333 895	76 599	243 511
廉　江	Lianjiang	1 543 443	546 518	522 537	474 388
雷　州	Leizhou	1 064 371	484 369	148 907	431 095
吴　川	Wuchuan	900 349	165 287	365 030	370 032
电　白	Dianbai	1 786 560	515 055	592 972	678 533

10-6 续表4 continued

沿海县 Coastal County		地区生产总值 Gross Regional Product	第一产业 Primary Industry	第二产业 Secondary Industry	第三产业 Tertiary Industry
惠 东	Huidong	2 166 368	241 468	1 125 363	799 537
海 丰	Haifeng	1 323 919	207 888	578 585	537 446
陆 丰	Lufeng	1 153 498	282 663	457 682	413 153
阳 西	Yangxi	720 668	304 531	229 461	186 676
阳 东	Yangdong	1 053 875	249 356	530 177	274 342
饶 平	Raoping	1 141 421	213 833	480 194	447 394
揭 东	Jiedong	1 842 597	250 775	1 077 325	514 497
惠 来	Huilai	1 033 795	285 016	473 633	275 146
广 西	**Guangxi**	**1 442 785**	**470 961**	**464 902**	**506 922**
合 浦	Hepu	1 113 432	399 784	369 151	344 497
东 兴	Dongxing	329 353	71 177	95 751	162 425
海 南	**Hainan**	**7 037 769**	**3 236 394**	**1 663 605**	**2 137 770**
琼 海	Qionghai	885 125	417 405	134 492	333 228
儋 州	Danzhou	1 092 846	611 168	130 683	350 995
文 昌	Wenchang	946 312	437 373	194 761	314 178
万 宁	Wanning	768 007	261 348	190 447	316 212
东 方	Dongfang	634 018	198 410	275 799	159 809
澄 迈	Chengmai	819 258	263 981	368 874	186 403
临 高	Lingao	578 008	407 789	39 345	130 874
昌 江	Changjiang	439 285	131 557	216 855	90 873
乐 东	Ledong	479 195	308 669	36 873	133 653
陵 水	Lingshui	395 715	198 694	75 476	121 545

注：本表各省数据为合计数，缺少上海崇明县，福建金门县数据。

Note: The data for the provinces are the totals, and lacking the data for Chongming of Shanghai Province; Jinmen of Fujian Province.

10-7 沿海地区财政收支
Financial Revenue and Expenditure by Coastal Regions

单位：万元 (10 000 yuan)

地 区 Region	一般预算收入 General Budgetary Revenue	一般预算支出 General Budgetary Expenditure
合 计 **Total**	**23 428.30**	**32 668.32**
天 津 Tianjin	1 068.81	1 376.84
河 北 Hebei	1 331.85	2 820.24
辽 宁 Liaoning	2 004.84	3 195.82
上 海 Shanghai	2 873.58	3 302.89
江 苏 Jiangsu	4 079.86	4 914.06
浙 江 Zhejiang	2 608.47	3 207.88
福 建 Fujian	1 151.49	1 695.09
山 东 Shandong	2 749.38	4 145.03
广 东 Guangdong	4 517.04	5 421.54
广 西 Guangxi	771.99	2 007.59
海 南 Hainan	270.99	581.34

10-8 沿海地区教育基本情况
Basic Conditions of Education by Coastal Regions

地　区 Region	高等学校数 （所） Institutions of Higher Education (unit)	本、专科在校学生数 （人） Number of Students Enrolled in Undergraduate or Specialized Courses (person)	本、专科毕（结）业生数 （人） Number of Students Graduated in Undergraduate or Specialized Courses (person)
全国总计 National Total	**2 358**	**22 317 929**	**5 754 245**
天　津 Tianjin	55	429 224	105 354
河　北 Hebei	110	1 105 118	297 092
辽　宁 Liaoning	112	880 247	219 564
上　海 Shanghai	67	515 661	133 716
江　苏 Jiangsu	150	1 649 430	478 868
浙　江 Zhejiang	101	884 867	233 741
福　建 Fujian	84	647 774	153 449
山　东 Shandong	132	1 631 373	444 003
广　东 Guangdong	131	1 426 624	334 187
广　西 Guangxi	70	567 516	138 089
海　南 Hainan	17	150 806	36 791

10-9 沿海地区卫生基本情况
Basic Conditions of Public Health by Coastal Regions

地　区 Region	卫生机构数 （个） Health Institutions (unit)	医疗机构床位数 （张） Total Beds (bed)	卫生机构人员 （人） Number of Employed Personnel in Health Institutions (person)
全国总计 **National Total**	**936 927**	**4 786 831**	**8 207 502**
天　津 Tianjin	4 542	48 828	96 732
河　北 Hebei	81 403	249 725	437 415
辽　宁 Liaoning	34 805	204 208	316 828
上　海 Shanghai	4 708	105 083	171 935
江　苏 Jiangsu	30 956	269 548	459 025
浙　江 Zhejiang	29 939	184 097	352 871
福　建 Fujian	27 017	113 043	199 519
山　东 Shandong	66 967	382 254	645 889
广　东 Guangdong	44 880	300 083	592 800
广　西 Guangxi	32 741	143 695	266 138
海　南 Hainan	4 678	25 981	51 985

10-10 沿海地区“十一五”节能目标完成情况
Indicators of Energy Consumption in the Eleventh Five-year Plan by Coastal Regions

地 区 Region	2005		2010	
	单位地区生产总值能耗（等价值）（吨标准煤/万元）Energy Consumption per Unit of GRP (ton of SCE/10 000 yuan)	“十一五”时期计划降低（±%）Targets During the Eleventh Five-year Plan (%)	单位地区生产总值能耗（等价值）（吨标准煤/万元）Energy Consumption per Unit of GRP (ton of SCE/10 000 yuan)	比2005年降低（%）Saving as Compared with 2005 (%)
天 津 Tianjin	1.046	20.00	0.826	21.00
河 北 Hebei	1.981	20.00	1.583	20.11
辽 宁 Liaoning	1.726	20.00	1.380	20.01
上 海 Shanghai	0.889	20.00	0.712	20.00
江 苏 Jiangsu	0.920	20.00	0.734	20.45
浙 江 Zhejiang	0.897	20.00	0.717	20.01
福 建 Fujian	0.937	16.00	0.783	16.45
山 东 Shandong	1.316	22.00	1.025	22.09
广 东 Guangdong	0.794	16.00	0.664	16.42
广 西 Guangxi	1.222	15.00	1.036	15.22
海 南 Hainan	0.920	12.00	0.808	12.14

注：地区生产总值和工业增加值按2005年价格计算。

Note: Gross regional product and industrial value-added are caculated at 2005 constant prices.

10-11 沿海地区用水情况
Water Use of Coastal Regions

地　区 Region	全年供水总量 （万立方米） Total Annual Volume of Water Supply (10 000 cu. m)	人均日生活用水量 （升） Per Capita Daily Consumption of Tap Water for Residential Use (liter)
全国总计 National Total	**5 078 745**	**171.4**
天　津 Tianjin	68 970	132.0
河　北 Hebei	166 430	123.0
辽　宁 Liaoning	261 879	121.0
上　海 Shanghai	336 637	174.8
江　苏 Jiangsu	482 821	220.4
浙　江 Zhejiang	270 044	185.4
福　建 Fujian	132 627	186.6
山　东 Shandong	290 866	129.5
广　东 Guangdong	806 144	250.0
广　西 Guangxi	147 291	249.7
海　南 Hainan	33 546	264.5

10-12 沿海地区全社会固定资产投资
Total Investment in Fixed Assets by the Whole Society of Coastal Regions

单位：亿元 (100 million yuan)

地　区 Region	全社会固定资产投资 Total Investment in Fixed Assets in the Whole Country	#农、林、牧、渔业 Agriculture, Forestry, Animal Husbandry and Fishery	#交通运输、仓储和邮政业 Transport, Storage and Post
全国总计 National Total	**278 121.9**	**7 923.1**	**30 074.5**
天　津 Tianjin	6 278.1	91.9	539.3
河　北 Hebei	15 083.4	565.2	1 521.2
辽　宁 Liaoning	16 043.0	358.3	1 080.8
上　海 Shanghai	5 108.9	16.4	655.2
江　苏 Jiangsu	23 184.3	221.9	1 162.1
浙　江 Zhejiang	12 376.0	99.6	1 068.7
福　建 Fujian	8 199.1	155.0	1 189.0
山　东 Shandong	23 280.5	628.3	1 362.3
广　东 Guangdong	15 623.7	219.0	1 820.0
广　西 Guangxi	7 057.6	242.4	842.5
海　南 Hainan	1 317.0	21.4	164.1

10-13 沿海地区按经营单位所在地分货物进出口总额
Total Value of Imports and Exports by Location of Importers/Exporters of Coastal Regions

单位：万美元 (10 000 US$)

地区 Region	进出口 Total	出口 Exports	进口 Imports
全国总计 **National Total**	**160 061 524**	**86 222 882**	**73 838 642**
天津 Tianjin	5 880 898	2 643 785	3 237 114
河北 Hebei	1 750 914	921 579	829 336
辽宁 Liaoning	3 899 198	2 063 953	1 835 245
上海 Shanghai	24 991 842	12 593 270	12 398 571
江苏 Jiangsu	34 720 361	19 231 357	15 489 004
浙江 Zhejiang	9 239 699	5 813 714	3 425 985
福建 Fujian	5 833 659	3 495 247	2 338 411
山东 Shandong	9 632 505	5 656 323	3 976 182
广东 Guangdong	48 449 160	28 184 708	20 264 452
广西 Guangxi	492 309	203 314	288 994
海南 Hainan	623 702	127 682	496 020

10-14 沿海地区城镇居民平均每人全年家庭收入和消费性支出
Per Capita Annual Income and Consumption Expenditure of Urban Households by Coastal Regions

单位：元 (yuan)

地 区 Region	总收入 Total Income	可支配收入 Disposable Income	消费性支出 Consumption Expenditure
全国平均水平 National Average	**21 033.42**	**19 109.44**	**13 471.45**
天 津 Tianjin	26 942.00	24 292.60	16 561.77
河 北 Hebei	17 334.42	16 263.43	10 318.32
辽 宁 Liaoning	20 014.57	17 712.58	13 280.04
上 海 Shanghai	35 738.51	31 838.08	23 200.40
江 苏 Jiangsu	25 115.40	22 944.26	14 357.49
浙 江 Zhejiang	30 134.79	27 359.02	17 858.20
福 建 Fujian	24 149.59	21 781.31	14 750.01
山 东 Shandong	21 736.94	19 945.83	13 118.24
广 东 Guangdong	26 896.86	23 897.80	18 489.53
广 西 Guangxi	18 742.21	17 063.89	11 490.08
海 南 Hainan	16 929.63	15 581.05	10 926.71

10-15 沿海地区农村居民家庭人均纯收入和生活消费支出
Per Capita Annual Net Income and Consumption Expenditure of Rural Households by Coastal Regions

单位：元 (yuan)

地　区 Region	纯收入 Net Income	生活消费支出 Consumption Expenditure
全国平均水平 **National Average**	**5 919.01**	**4 381.82**
天　津 Tianjin	10 074.86	4 936.73
河　北 Hebei	5 957.98	3 844.92
辽　宁 Liaoning	6 907.93	4 489.50
上　海 Shanghai	13 977.96	10 210.46
江　苏 Jiangsu	9 118.24	6 542.87
浙　江 Zhejiang	11 302.55	8 928.89
福　建 Fujian	7 426.86	5 498.33
山　东 Shandong	6 990.28	4 807.18
广　东 Guangdong	7 890.25	5 515.58
广　西 Guangxi	4 543.41	3 455.29
海　南 Hainan	5 275.37	3 446.24

10-16 沿海地区人口情况
Population of Coastal Regions

单位：万人 (10 000 persons)

地　区 Region	年末总人口 Total Population by the End of the Year
全国总计 **National Total**	**134 091**
天　津 Tianjin	1 299
河　北 Hebei	7 194
辽　宁 Liaoning	4 375
上　海 Shanghai	2 303
江　苏 Jiangsu	7 869
浙　江 Zhejiang	5 447
福　建 Fujian	3 693
山　东 Shandong	9 588
广　东 Guangdong	10 441
广　西 Guangxi	4 610
海　南 Hainan	869

注:1. 本表数据根据2010年第六次全国人口普查数据推算。

2. 全国总人口包括现役军人和难以确定常住地人口，分地区数据未包括。

Note: 1. The data are calculated from the census statistics in 2010.

2. Total population include servicemen and those population with permanent residence difficult to define while regional breakdowns do not include them.

10-17 沿海城市人口情况（2009年）
Population of Coastal Cities, 2009

单位：万人　　(10 000 persons)

沿海城市 Coastal City		年末总人口 Total Population by the End of the Year
合　计	**Total**	**24 262.5**
天　津	**Tianjin**	**1 228.2**
河　北	**Hebei**	**1 738.6**
唐　山	Tangshan	733.9
秦皇岛	Qinhuangdao	287.2
沧　州	Cangzhou	717.5
辽　宁	**Liaoning**	**1 784.9**
大　连	Dalian	584.8
丹　东	Dandong	242.6
锦　州	Jinzhou	310.2
营　口	Yingkou	235.0
盘　锦	Panjin	130.0
葫芦岛	Huludao	282.3
上　海	**Shanghai**	**1 400.7**
江　苏	**Jiangsu**	**2 065.7**
南　通	Nantong	762.7
连云港	Lianyungang	490.6
盐　城	Yancheng	812.4

10-17 续表1 continued

沿海城市 Coastal City		年末总人口 Total Population by the End of the Year
浙　江	**Zhejiang**	**3 486.1**
杭　州	Hangzhou	683.4
宁　波	Ningbo	571.0
温　州	Wenzhou	779.1
嘉　兴	Jiaxing	339.6
绍　兴	Shaoxing	437.7
舟　山	Zhoushan	96.8
台　州	Taizhou	578.5
福　建	**Fujian**	**2 624.1**
福　州	Fuzhou	637.9
厦　门	Xiamen	177.0
莆　田	Putian	319.6
泉　州	Quanzhou	680.8
漳　州	Zhangzhou	471.8
宁　德	Ningde	337.0
山　东	**Shandong**	**3 383.7**
青　岛	Qingdao	762.9
东　营	Dongying	184.6
烟　台	Yantai	652.0
潍　坊	Weifang	867.9
威　海	Weihai	253.0
日　照	Rizhao	285.8
滨　州	Binzhou	377.5

10-17 续表2 continued

沿海城市	Coastal City	年末总人口 Total Population by the End of the Year
广　东	**Guangdong**	**5 718.2**
广　州	Guangzhou	794.6
深　圳	Shenzhen	246.0
珠　海	Zhuhai	102.7
汕　头	Shantou	510.7
江　门	Jiangmen	391.5
湛　江	Zhanjiang	763.1
茂　名	Maoming	735.3
惠　州	Huizhou	324.4
汕　尾	Shanwei	340.6
阳　江	Yangjiang	275.7
东　莞	Dongguan	178.7
中　山	Zhongshan	147.9
潮　州	Chaozhou	257.9
揭　阳	Jieyang	649.1
广　西	**Guangxi**	**618.3**
北　海	Beihai	160.2
防城港	Fangchenggang	86.9
钦　州	Qinzhou	371.2
海　南	**Hainan**	**214.0**
海　口	Haikou	158.3
三　亚	Sanya	55.7

注：本表各省数据为合计数。

Note: The data for the provinces are the totals.

10-18 沿海县人口情况（2009年）
Population of Coastal Counties, 2009

单位：万人 (10 000 persons)

沿海县	Coastal County	年末总人口 Total Population by the End of the Year
合　计	**Total**	**8 527.62**
河　北	**Hebei**	**297.00**
滦　南	Luannan	58.29
乐　亭	Leting	49.69
唐　海	Tanghai	14.22
昌　黎	Changli	55.59
抚　宁	Funing	52.60
黄　骅	Huanghua	43.85
海　兴	Haixing	22.76
辽　宁	**Liaoning**	**659.80**
长　海	Changhai	7.40
瓦房店	Wafangdian	102.60
普兰店	Pulandian	81.90
庄　河	Zhuanghe	90.80
东　港	Donggang	61.20
凌　海	Linghai	53.30
盖　州	Gaizhou	73.20
大　洼	Dawa	39.70
盘　山	Panshan	29.60
绥　中	Suizhong	64.50
兴　城	Xingcheng	55.60
上　海	**Shanghai**	
崇　明	Chongming	
江　苏	**Jiangsu**	**1 272.19**

10-18 续表1 continued

沿海县 Coastal County		年末总人口 Total Population by the End of the Year
海　安	Hai'an	93.68
如　东	Rudong	105.30
启　东	Qidong	111.58
海　门	Haimen	99.84
赣　榆	Ganyu	110.80
东　海	Donghai	113.16
灌　云	Guanyun	101.52
灌　南	Guannan	76.47
响　水	Xiangshui	60.93
滨　海	Binhai	116.15
射　阳	Sheyang	96.58
东　台	Dongtai	113.77
大　丰	Dafeng	72.41
浙　江	**Zhejiang**	**1 462.40**
象　山	Xiangshan	53.71
宁　海	Ninghai	60.50
余　姚	Yuyao	83.25
慈　溪	Cixi	103.52
奉　化	Fenghua	48.21
洞　头	Dongtou	12.74
平　阳	Pingyang	86.26
苍　南	Cangnan	127.59
瑞　安	Rui'an	118.75
乐　清	Yueqing	122.49
海　盐	Haiyan	37.07
海　宁	Haining	65.50
平　湖	Pinghu	48.51
绍　兴	Shaoxing	71.78
上　虞	Shangyu	77.42

10-18 续表2 continued

沿海县 Coastal County		年末总人口 Total Population by the End of the Year
岱　山	Daishan	19.14
嵊　泗	Shengsi	7.97
玉　环	Yuhuan	41.47
三　门	Sanmen	42.60
温　岭	Wenling	118.45
临　海	Linhai	115.47
福　建	**Fujian**	**1 271.96**
连　江	Lianjiang	62.51
罗　源	Luoyuan	25.46
平　潭	Pingtan	38.97
福　清	Fuqing	125.18
长　乐	Changle	67.37
仙　游	Xianyou	106.64
惠　安	Hui'an	95.38
石　狮	Shishi	31.49
晋　江	Jinjiang	105.69
南　安	Nan'an	150.13
云　霄	Yunxiao	43.09
漳　浦	Zhangpu	84.31
诏　安	Zhao'an	59.29
东　山	Dongshan	20.73
龙　海	Longhai	80.96
霞　浦	Xiapu	52.41
福　安	Fu'an	64.76
福　鼎	Fuding	57.59
山　东	**Shandong**	**1 221.27**
胶　州	Jiaozhou	80.03
即　墨	Jimo	112.61

10-18 续表3 continued

沿海县 Coastal County		年末总人口 Total Population by the End of the Year
胶　南	Jiaonan	83.74
垦　利	Kenli	21.95
利　津	Lijin	29.81
广　饶	Guangrao	49.54
长　岛	Changdao	4.32
龙　口	Longkou	63.30
莱　阳	Laiyang	87.61
莱　州	Laizhou	85.91
蓬　莱	Penglai	44.95
招　远	Zhaoyuan	57.06
海　阳	Haiyang	66.66
寿　光	Shouguang	103.28
昌　邑	Changyi	58.07
文　登	Wendeng	64.26
荣　成	Rongcheng	66.85
乳　山	Rushan	57.32
无　棣	Wudi	45.08
沾　化	Zhanhua	38.92
广　东	**Guangdong**	**1 702.66**
南　澳	Nan'ao	7.30
台　山	Taishan	98.59
恩　平	Enping	50.14
遂　溪	Suixi	103.90
徐　闻	Xuwen	72.43
廉　江	Lianjiang	163.58
雷　州	Leizhou	163.15
吴　川	Wuchuan	108.28
电　白	Dianbai	140.94

10-18 续表4 continued

沿海县 Coastal County		年末总人口 Total Population by the End of the Year
惠　东	Huidong	80.42
海　丰	Haifeng	83.79
陆　丰	Lufeng	173.49
阳　西	Yangxi	50.51
阳　东	Yangdong	47.61
饶　平	Raoping	100.84
揭　东	Jiedong	127.19
惠　来	Huilai	130.50
广　西	**Guangxi**	**112.25**
东　兴	Dongxing	12.49
合　浦	Hepu	99.76
海　南	**Hainan**	**528.09**
琼　海	Qionghai	48.83
儋　州	Danzhou	100.83
文　昌	Wenchang	57.68
万　宁	Wanning	60.10
东　方	Dongfang	43.49
澄　迈	Chengmai	53.75
临　高	Lingao	48.40
昌　江	Changjiang	26.03
乐　东	Ledong	52.58
陵　水	Lingshui	36.40

注：本表各省数据为合计数，缺少福建金门县数据。

Note: The data for the provinces are the totals, and lacking the data for Jinmen of Fujian Province.

10-19 沿海地区就业人员情况
Number of Employed Persons by Coastal Regions

单位：万人 (10 000 persons)

地 区 Region	就业人员 Number of Employed Persons	城镇就业人员 Urban Employed Persons
全国总计 National Total	**76 105.0**	**34 687.0**
天 津 Tianjin	520.8	329.0
河 北 Hebei	3 790.2	813.6
辽 宁 Liaoning	2 238.1	1 029.6
上 海 Shanghai	924.7	736.0
江 苏 Jiangsu	4 731.7	2 061.1
浙 江 Zhejiang	3 989.2	1 642.4
福 建 Fujian	2 181.3	785.5
山 东 Shandong	5 654.7	1 592.8
广 东 Guangdong	5 776.9	2 351.7
广 西 Guangxi	2 945.3	558.1
海 南 Hainan	445.7	161.1

10-20 沿海城市就业人员情况（2009年）

Number of Employed Persons by Coastal Cities, 2009

单位：万人 (10 000 persons)

沿海城市 Coastal City	就业人员 Number of Employed Persons	城镇就业人员 Urban Employed Persons
合 计 Total	**15 942.70**	**6 680.90**
天 津 Tianjin	**677.10**	**505.80**
河 北 Hebei	**986.90**	**285.50**
唐 山 Tangshan	432.20	146.50
秦皇岛 Qinhuangdao	163.90	55.30
沧 州 Cangzhou	390.80	83.70
辽 宁 Liaoning	**1 049.80**	**527.10**
大 连 Dalian	383.30	243.30
丹 东 Dandong	118.20	40.00
锦 州 Jinzhou	167.40	66.30
营 口 Yingkou	136.40	63.20
盘 锦 Panjin	107.60	71.00
葫芦岛 Huludao	136.90	43.30
上 海 Shanghai	**1 064.40**	**385.40**
江 苏 Jiangsu	**1 092.80**	**314.00**
南 通 Nantong	460.50	117.00
连云港 Lianyungang	290.30	66.50
盐 城 Yancheng	342.00	130.50
浙 江 Zhejiang	**2 670.00**	**1 324.00**
杭 州 Hangzhou	597.50	383.90
宁 波 Ningbo	443.90	263.30
温 州 Wenzhou	546.90	242.00
嘉 兴 Jiaxing	308.10	165.10
绍 兴 Shaoxing	329.00	118.30
舟 山 Zhoushan	66.00	30.30
台 州 Taizhou	378.60	121.10
福 建 Fujian	**1 686.30**	**664.80**
福 州 Fuzhou	367.60	150.90
厦 门 Xiamen	216.50	194.80
莆 田 Putian	170.70	39.10

10-20 续表 continued

沿海城市 Coastal City	就业人员 Number of Employed Persons	城镇就业人员 Urban Employed Persons
泉　州　Quanzhou	505.60	194.20
漳　州　Zhangzhou	263.50	54.40
宁　德　Ningde	162.40	31.40
山　东　Shandong	**2 063.00**	**695.40**
青　岛　Qingdao	501.50	226.80
东　营　Dongying	110.40	50.50
烟　台　Yantai	400.60	142.40
潍　坊　Weifang	501.60	133.30
威　海　Weihai	150.40	61.80
日　照　Rizhao	170.60	34.00
滨　州　Binzhou	227.90	46.60
广　东　Guangdong	**4 236.70**	**1 851.00**
广　州　Guangzhou	738.70	375.30
深　圳　Shenzhen	692.50	454.00
珠　海　Zhuhai	98.10	84.90
汕　头　Shantou	249.90	108.00
江　门　Jiangmen	233.60	106.40
湛　江　Zhanjiang	312.50	73.60
茂　名　Maoming	347.40	59.00
惠　州　Huizhou	245.50	161.80
汕　尾　Shanwei	120.00	51.00
阳　江　Yangjiang	159.90	43.10
东　莞　Dongguan	429.10	96.50
中　山　Zhongshan	210.00	120.20
潮　州　Chaozhou	139.90	21.80
揭　阳　Jieyang	259.60	95.40
广　西　Guangxi	**284.80**	**46.00**
北　海　Beihai		9.10
防城港　Fangchenggang	55.30	12.70
钦　州　Qinzhou	229.50	24.20
海　南　Hainan	**130.90**	**81.90**
海　口　Haikou	103.00	69.40
三　亚　Sanya	27.90	12.50

注：本表各省数据为合计数。

Note: The data for the provinces are the totals.

10-21 沿海县就业人员情况（2009年）
Number of Employed Persons of Coastal Counties, 2009

单位：人 (person)

沿海县 Coastal County		城镇单位职工人数 Urban Employed Persons	乡村从业人员 Rural Laborer
合 计	**Total**	**6 361 741**	**37 920 849**
河 北	**Hebei**	**182 924**	**1 420 614**
滦 南	Luannan	26 977	288 634
乐 亭	Leting	21 679	265 001
唐 海	Tanghai	40 923	68 128
昌 黎	Changli	21 515	280 314
抚 宁	Funing	30 444	244 881
黄 骅	Huanghua	30 670	170 824
海 兴	Haixing	10 716	102 832
辽 宁	**Liaoning**	**259 561**	**2 669 360**
长 海	Changhai	7 099	31 335
瓦房店	Wafangdian	38 882	374 223
普兰店	Pulandian	37 561	308 534
庄 河	Zhuanghe	36 863	349 430
东 港	Donggang	25 631	264 261
凌 海	Linghai	20 309	225 502
盖 州	Gaizhou	16 631	320 440
大 洼	Dawa	23 923	168 605
盘 山	Panshan	20 138	166 254
绥 中	Suizhong	15 418	270 556
兴 城	Xingcheng	17 106	190 220
江 苏	**Jiangsu**	**633 339**	**5 289 400**
海 安	Hai'an	65 498	398 100

10-21 续表1 continued

沿海县 Coastal County	城镇单位职工人数 Urban Employed Persons	乡村从业人员 Rural Laborer
如 东 Rudong	62 102	494 300
启 东 Qidong	57 458	565 800
通 州 Tongzhou		
海 门 Haimen	65 995	503 500
赣 榆 Ganyu	34 430	420 500
东 海 Donghai	38 304	461 400
灌 云 Guanyun	39 309	375 300
灌 南 Guannan	31 843	313 900
响 水 Xiangshui	30 124	210 500
滨 海 Binhai	35 519	413 500
射 阳 Sheyang	58 789	340 500
东 台 Dongtai	60 541	478 400
大 丰 Dafeng	53 427	313 700
浙 江 Zhejiang	**1 833 500**	**7 981 600**
象 山 Xiangshan	217 400	270 300
宁 海 Ninghai	50 100	331 700
余 姚 Yuyao	63 000	463 300
慈 溪 Cixi	83 200	778 500
奉 化 Fenghua	58 100	262 100
洞 头 Dongtou	7 602	53 100
平 阳 Pingyang	66 679	388 100
苍 南 Cangnan	72 374	657 400
瑞 安 Rui'an	121 446	632 900
乐 清 Yueqing	213 233	666 000
海 盐 Haiyan	63 088	210 000
海 宁 Haining	121 756	312 600
平 湖 Pinghu	127 095	217 000
绍 兴 Shaoxing	177 699	465 600
上 虞 Shangyu	117 137	373 000

10-21 续表2 continued

沿海县 Coastal County	城镇单位职工人数 Urban Employed Persons	乡村从业人员 Rural Laborer
岱 山 Daishan	17 467	85 500
嵊 泗 Shengsi	10 500	27 100
玉 环 Yuhuan	45 724	340 900
三 门 Sanmen	25 436	220 200
温 岭 Wenling	81 119	651 700
临 海 Linhai	93 345	574 600
福 建 Fujian	**1 439 722**	**5 897 468**
连 江 Lianjiang	35 310	293 575
罗 源 Luoyuan	13 797	99 816
平 潭 Pingtan	15 388	185 168
福 清 Fuqing	188 352	543 352
长 乐 Changle	37 133	264 334
仙 游 Xianyou	41 267	491 685
惠 安 Hui'an	210 315	463 838
石 狮 Shishi	72 077	125 978
晋 江 Jinjiang	512 685	622 912
南 安 Nan'an	81 923	775 552
云 霄 Yunxiao	16 798	163 526
漳 浦 Zhangpu	44 600	427 068
诏 安 Zhao'an	20 648	309 733
东 山 Dongshan	17 055	79 324
龙 海 Longhai	68 058	372 518
霞 浦 Xiapu	15 790	219 207
福 安 Fu'an	27 397	198 198
福 鼎 Fuding	21 129	261 684
山 东 Shandong	**1 128 137**	**5 508 252**
胶 州 Jiaozhou	128 360	366 120

10-21 续表3 continued

沿海县 Coastal County	城镇单位职工人数 Urban Employed Persons	乡村从业人员 Rural Laborer
即　墨 Jimo	130 391	567 699
胶　南 Jiaonan	124 033	353 602
垦　利 Kenli	23 824	82 604
利　津 Lijin	15 114	139 664
广　饶 Guangrao	58 524	257 516
长　岛 Changdao	4 927	14 227
龙　口 Longkou	65 277	266 806
莱　阳 Laiyang	66 710	421 996
莱　州 Laizhou	66 861	369 669
蓬　莱 Penglai	40 374	202 296
招　远 Zhaoyuan	54 948	217 374
寿　光 Shouguang	64 341	406 985
海　阳 Haiyang	32 238	354 646
昌　邑 Changyi	28 441	261 922
文　登 Wendeng	64 423	271 534
荣　成 Rongcheng	84 650	219 565
乳　山 Rushan	40 358	278 540
无　棣 Wudi	20 152	246 777
沾　化 Zhanhua	14 191	208 710
广　东 Guangdong	**645 525**	**6 823 987**
南　澳 Nan'ao	5 054	24 397
台　山 Taishan	46 845	491 821
恩　平 Enping	33 277	171 897
遂　溪 Suixi	38 717	429 780
徐　闻 Xuwen	36 872	309 979
廉　江 Lianjiang	56 793	337 459
雷　州 Leizhou	55 194	659 638
吴　川 Wuchuan	36 193	521 968

10-21 续表4 continued

沿海县 Coastal County	城镇单位职工人数 Urban Employed Persons	乡村从业人员 Rural Laborer
电 白 Dianbai	50 578	638 972
惠 东 Huidong	58 786	366 324
海 丰 Haifeng	33 113	390 804
陆 丰 Lufeng	43 216	572 204
阳 西 Yangxi	23 337	255 871
阳 东 Yangdong	28 800	255 576
饶 平 Raoping	27 952	425 962
揭 东 Jiedong	32 001	594 117
惠 来 Huilai	38 797	377 218
广 西 Guangxi	**43 984**	**401 302**
东 兴 Dongxing	6 952	54 400
合 浦 Hepu	37 032	346 902
海 南 Hainan	**195 049**	**1 928 866**
琼 海 Qionghai	23 090	187 370
儋 州 Danzhou	27 091	324 511
文 昌 Wenchang	24 192	218 306
万 宁 Wanning	21 066	182 018
东 方 Dongfang	19 591	171 967
澄 迈 Chengmai	22 807	200 099
临 高 Lingao	13 345	187 592
昌 江 Changjiang	17 597	85 836
乐 东 Ledong	16 212	234 845
陵 水 Lingshui	10 058	136 322

注：本表各省数据为合计数，缺少福建金门县数据。

Note: The data for the provinces are the totals, and lacking the data for Jinmen of Fujian Province.

主要统计指标解释

1. 国内(或地区)生产总值 指一个国家（或地区）所有常驻单位在一定时期内生产活动的最终成果。国内生产总值有三种表现形态，即价值形态、收入形态和产品形态。从价值形态看，它是所有常驻单位在一定时期内生产的全部货物和服务价值超过同期中间投入的全部非固定资产货物和服务价值的差额，即所有常驻单位的增加值之和；从收入形态看，它是所有常驻单位在一定时期内创造并分配给常驻单位和非常驻单位的初次收入分配之和；从产品形态看，它是所有常驻单位在一定时期内最终使用的货物和服务价值与货物和服务净出口价值之和。在实际核算中，国内生产总值有三种计算方法，即生产法、收入法和支出法。三种方法分别从不同的方面反映国内生产总值及其构成。

2. 三次产业 是根据社会生产活动历史发展的顺序对产业结构的划分，产品直接取自自然界的部门称为第一产业，对初级产品进行再加工的部门称为第二产业，为生产和消费提供各种服务的部门称为第三产业。它是世界上较为通用的产业结构分类，但各国的划分不尽一致。我国的三次产业划分是：

第一产业：是指农、林、牧、渔业。

第二产业：是指采矿业，制造业，电力、燃气及水的生产和供应业，建筑业。

第三产业：是指除第一、二产业以外的其他行业。第三产业包括：交通运输、仓储和邮政业，信息传输、计算机服务和软件业，批发和零售业，住宿和餐饮业，金融业，房地产业，租赁和商务服务业，科学研究、技术服务和地质勘查业，水利、环境和公共设施管理业，居民服务和其他服务业，教育，卫生、社会保障和社会福利业，文化、体育和娱乐业，公共管理和社会组织，国际组织。

3. 增加值 是指各行各业生产经营和劳务活动的最终成果，采用生产法和收入法两种方法计算。

生产法 是从货物和服务活动在生产过程中形成的总产品入手，剔除生产过程中投入的中间产品价值，得到新增价值的方法。

收入法 又称分配法。按收入法计算国内生产总值是从生产过程创造的收入的角度对常驻单位的生产活动成果进行核算；按照此法计算，增加值由劳动者报酬、固定资产折旧、生产税净额和营业盈余四个部分组成。

4. 年末总人口 是指每年 12 月 31 日 24 时一定地区范围内的有生命的个人的人口总和。

5. 财政收入 指国家财政参与社会产品分配所取得的收入，是实现国家职能的财力保证。财政收入所包括的内容几经变化，目前主要包括：

(1)各项税收：包括增值税、营业税、消费税、土地增值税、城市维护建设税、资源税、城市土地使用税、企业所得税、个人所得税、关税、证券交易印花税、车辆购置税、农牧业税和耕地占用税等。

(2)专项收入：包括排污费收入、城市水资源费收入、矿产资源补偿费收入、教育费附加收入等。

(3)其他收入：包括利息收入、基本建设贷款归还收入、基本建设收入、捐赠收入等。

(4)国有企业亏损补贴：此项为负收入，冲减财政收入。主要包括对工业企业、商业企业、粮食企业的补贴。

6. 财政支出 国家财政将筹集起来的资金进行分配使用，以满足经济建设和各项事业的需要。

7. 基本建设支出 指按国家有关规定，属于基本建设范围内的基本建设有偿使用、拨款、资本金支出以及经国家批准对专项和政策性基建投资贷款，在部门的基建投资额中统筹支付的贴息支出。

8. 普通高等学校 指按照国家规定的设置标准和审批程序批准举办的，通过全国普通高等学校统一招生考试，招收高中毕业生为主要培养对象，实施高等教育的全日制大学、独立设置的学院和高等专科学校、高等职业学校和其他机构。

大学、独立设置的学院主要实施本科层次以上教育，高等专科学校、高等职业学校实施专科层次教育，其他机构是承担国家普通招生计划任务不计校数的机构。包括普通高等学校分校和批准筹建的普通高等学校等。

9. 卫生机构 包括医疗机构、疾病预防控制中心(防疫站)、采供血机构、卫生监督及监测(检验)机构、医学科研和在职培训机构、健康教育所等。

10. 医疗机构 包括医院、社区卫生服务中心(站)、疗养院、卫生院、门诊部、诊所(卫生所、医务室)、妇幼保健院(所、站)、专科疾病防治院(所、站)、急救中心(站)和临床检验中心。医疗机构分为非赢利性医疗机构和赢利性医疗机构。

11. 单位国内生产总值能耗 指一定时期内，一个国家或地区每生产一个单位的国内生产总值所消耗的能源。计算公式为：

$$\text{单位国内生产总值能源}=\frac{\text{能源消费总量}}{\text{国内生产总值}}$$

12. 单位国内生产总值电耗 指一定时期内，一个国家或地区每生产一个单位的国内生产总值所消耗的电力。计算公式为：

$$\text{单位国内生产总值电耗}=\frac{\text{全社会用电量}}{\text{国内生产总值}}$$

13. 单位工业增加值能耗 指一定时期内，一个国家或地区每生产一个单位的工业增加值所消耗的能源。计算公式为：

$$\text{单位工业增加值能耗}=\frac{\text{工业能源消费总量}}{\text{工业增加值}}$$

14. 用水总量 指分配给各类用户的包括输水损失在内的毛用水量之和，不包括海水直接利用量。

15. 全社会固定资产投资 是以货币形式表现的在一定时期内全社会建造和购置固定资产的工作量以及与此有关的费用的总称。该指标是反映固定资产投资规模、结构和发展速度的综合性指标，又是观察工程进度和考核投资效果的重要依据。全社会固定资产投资按登记注册类型可分为国有、集体、个体、联营、股份制、外商、港澳台商、其他等。

16. 进出口总额 指实际进出我国国境的货物总金额。包括对外贸易实际进出口货物，来料加工装配进出口货物，国家间、联合国及国际组织无偿援助物资和赠送品，华侨、港澳台同胞和外籍华人捐赠品，租赁期满归承租人所有的租赁货物，进料加工进出口货物，边境地方贸易及边境地区小额贸易进出口货物(边民互市贸易除外)，中外合资企业、中外合作经营企业、外商独资经营企业进出口货物和公用物品，到、离岸价格在规定限额以上的进出口货样和广告品(无商业价值、无使用价值和免费提供出口的除外)，从保税仓库提取在中国境内销售的进口货物，以及其他进出口货物。该指标可以观察一个国家在对外贸易方面的总规模。我国规定出口货物按离岸价格统计，进口货物按到岸价格统计。

17. 商品经营单位所在地进、出口额　指在所在地海关注册登记的有进出口经营权的企业实际进、出口额。

18. 城镇家庭可支配收入　指家庭成员得到可用于最终消费支出和其它非义务性支出以及储蓄的总和，即居民家庭可以用来自由支配的收入。它是家庭总收入扣除交纳的所得税、个人交纳的社会保障支出以及记账补贴后的收入。计算公式为：

可支配收入=家庭总收入-交纳所得税-个人交纳的社会保障支出-记账补贴

19. 城镇家庭消费性支出　指家庭用于日常生活的支出，包括食品、衣着、家庭设备用品及服务、医疗保健、交通和通信、娱乐教育文化服务、居住、杂项商品和服务等八大类支出。

20. 总收入　指调查期内农村住户和住户成员从各种来源渠道得到的收入总和。按收入的性质划分为工资性收入、家庭经营收入、财产性收入和转移性收入。

21. 纯收入　指农村住户当年从各个来源得到的总收入相应地扣除所发生的费用后的收入总和。计算方法：

纯收入=总收入-税费支出-家庭经营费用支出-生产性固定资产折旧-赠送农村亲友支出

纯收入主要用于再生产投入和当年生活消费支出，也可用于储蓄和各种非义务性支出。“农民人均纯收入”按人口平均的纯收入水平，反映的是一个地区或一个农户农村居民的平均收入水平。

22. 人口数　指一定时点、一定地区范围内有生命的个人总和。

年度统计的年末人口数指每年 12 月 31 日 24 时的人口数。年度统计的全国人口总数内未包括香港、澳门特别行政区和台湾省以及海外华侨人数。

23. 城镇人口　城镇人口是指居住在城镇范围内的全部常住人口

24. 就业人员　指在 16 周岁及以上，从事一定社会劳动并取得劳动报酬或经营收入的人员。这一指标反映了一定时期内全部劳动力资源的实际利用情况，是研究我国基本国情国力的重要指标。

Explanatory Notes on Main Statistical Indicators

1. Gross Domestic Product (GDP) refers to the final result of the primary distribution of the income created by all the resident units of a country (or a region) during a certain period of time. Gross domestic product is expressed in three different forms, i.e. value, income, and products respectively. The form of value refers to the total value of all products and services produced by all resident units during a certain period of time minus total value of intermediate input of materials and services of the nature of non-fixed assets or the summation of the value added of all resident units: the form of income includes all the income created by all resident units and distributed primarily to all resident and non-resident units; the form of products refers to the summation of the value of the products and services finally used and the net export value of products and services by all resident units during a given period of time. In the practice of national accounting, gross domestic product is calculated with three approaches, i.e. production approach, income approach, and expenditure approach, which reflect the gross domestic product and its composition from different aspects.

2. Three Industries Industrial structure is classified according to the sequence of historical development of social productive activities. Primary industry refers to the extraction of natural resources; secondary industry involves processing of primary products; and tertiary industry provides services of various kinds for production and consumption. The above classification is universal in the world although it varies to some extent from country to country. The three industries in China are divided as follows:

Primary industry: refers to farming, forestry, sideline production and fishery.

Secondary industry: refers to such industries as mining, manufacturing, production and supply of electric power, fuel gas and water, and construction.

Tertiary industry: refers to all the other industries not included in the primary or secondary industries. It includes such industries as communications and transportation, storage and postal service; information transmission, computer service and software; wholesale and retailing; accommodation and catering; financial service; real estate; charter business and commercial affairs service; scientific research, technological service and geological survey; water conservancy, environmental and other public facilities management; residents service and other service trades; education, health, social security and social welfare; culture, sports and entertainment business as well as public administration and social organizations, and international organizations.

3. Added Value refers to the final result of production operation and labor activities of all trades and professions, which is calculated by using the methods of production and income.

Production Method: refers to the method whereby to get the newly added value by proceeding from the gross product of goods and service activities occurring in the course of production and then rejecting the value of intermediate product input in the course of production.

Income Method: is also called the distribution method. The calculation of the gross domestic product (GDP) by the income method is the accounting of the result of productive activities of permanent units from the angle of the income created in the course of production. According to this method, the added value is composed of the payment for laborers, depreciation for fixed assets, net tax on production and business surplus.

4. Total Population by the End of the Year refers to the sum of living individuals within a particular range of area at 24:00 on December 31 of each year.

5. Government Revenue refers to the income obtained by the government finance through participating in the distribution of social products. It is the financial guarantee to ensure government functioning. The contents of government revenue have changed several times. Now it includes the following main items:

(1) Various tax revenues, including value added tax, business tax, consumption tax, land

value-added tax, tax on city maintenance and construction, resources tax, tax on use of urban land, enterprise income tax, personal income tax, tariff, stamp tax on security transactions, tax on purchase of motor vehicles, tax on agriculture and animal husbandry and tax on occupancy of cultivated land, etc.

(2) Special revenues, including revenues from the fee on sewage treatment, fee on urban water resources, fee for the compensation of mineral resources and extra-charges for education, etc.

(3) Other revenues, including revenues from interest, repayment of capital construction loan, capital construction projects, and donations and grants.

(4) Subsidies for the losses of State-owned enterprises. This is an item of negative revenue, counteracting revenues and consisting of subsidies to industrial, commercial and grain purchasing and supply enterprises.

6. Government Expenditure refers to the distribution and use of the funds which the government finance has raised, so as to meet the needs of economic construction and various causes.

7. Expenditure for capital construction It refers to the non-gratuitous use of, appropriation of funds for and capital outlay on capital construction in the area of capital construction. It also covers the loans on capital construction approved by the government for special purposes or policy purposes and the expenditure with discount paid in an overall way within the amount of the funds appropriated to the departments for capital construction.

8. Regular Institutions of Higher Learning refer to educational establishments set up according to the government evaluation and approval procedures, enrolling graduates from senior secondary schools and providing higher education courses and training for senior professionals. They include full-time universities, colleges, institutions of higher professional education, institutions of higher vocational education and others.

Universities and colleges primarily provide undergraduate courses; institutions of higher professional education and institutions of higher vocational education primarily provide professional trainings; and others refer to educational establishments, which are responsible for enrolling higher education students under the State Plan but not enumerated in the total number of schools, including: branch schools of universities and colleges, and universities and colleges that have been approved and under plan for construction.

9. Health Care Institutions include: medical institutions, disease prevention and control centres (epidemic prevention stations), blood gathering and supplying institutions, health supervision and inspection (check up) institutions, medicinal scientific research and on-job training institutions, health education centres and so on.

10. Medical Organizations include: hospitals, health service centres (stations) in communities, sanatoria, health centres, out-patient clinics, clinics (health stations and infirmaries), maternity and child care agencies (centres and stations), special disease prevention and curing agencies (centres and stations), first aid centres (stations) and clinical inspection centres. Medical organizations are grouped by two types: profit-making and non-profit-making medical organizations.

11. Energy Consumption per Unit of GDP refers to the energy consumption per unit of Gross Domestic Product in a country or the Gross Regional Product in a region in the same reference period. The formula is:

$$\text{Energy Consumption per Unit of GDP} = \frac{\text{Total Energy Consumption}}{\text{Gross Domestic Product}}$$

12. Electricity Consumption per Unit of GDP refers to the electricity consumption per unit of Gross Domestic Product in a country or the Gross Regional Product in a region in the same reference period. The formula is:

$$\text{Electricity Consumption per Unit of GDP} = \frac{\text{Total Electricity Consumption}}{\text{Gross Domestic Product}}$$

13. Energy Consumption per Unit of Industrial Value-added refers to the energy consumption per unit of industrial value-added in a country or region in the same reference period. The formula is:

$$\text{Energy Consumption per Unit of Industrial Value-added} = \frac{\text{Industry Energy Consumption}}{\text{Industrial Value-added.}}$$

14. Gross Amount of Water Used refers to gross water use distributed to users, including loss during transportation, broken down into use by agriculture, industry, living consumption and ecological protection.

15. Total Investment in Fixed Assets in the Whole Country refers to the volume of activities in construction and purchases of fixed assets of the whole country and related fees, expressed in monetary terms during the reference period. It is a comprehensive indicator which shows the size, structure and growth of the investment in fixed assets, providing a basis for observing the progress of construction projects and evaluating results of investment. Total investment in fixed assets in the whole country includes, by type of ownership, the investment by State-owned units, collective-owned units, individuals, joint ownership units, share-holding units, as well as investments by entrepreneurs from foreign countries and from Hong Kong, Macao and Taiwan, and by other units.

16. Total Imports and Exports at Customs refer to the real value of commodities imported and exported across the border of China. They include the actual imports and exports through foreign trade, imported and exported goods under the processing and assembling trades and materials, supplies and gifts as aid given gratis between governments and by the United Nations and other international organizations, and contributions donated by overseas Chinese compatriots in Hong Kong and Macao and Chinese with foreign citizenship, leasing commodities owned by tenant at the expiration of leasing period, the imported and exported commodities processed with imported materials, commodities trading in border areas (excluding mutual exchange goods), the imported and exported commodities and articles for public use of the Sino-foreign joint ventures, cooperative enterprises and ventures with sole foreign investment. Also included is the import or export of samples and advertising goods for which the CIF or FOB value is beyond the permitted ceiling (excluding goods of no trading or use value and free commodities for export), imported goods sold in China from bonded warehouses and other imported or exported goods. The indicator of the total imports and exports at customs can be used to observe the total size of external trade in a country. In accordance with the stipulation of the Chinese government, imports are calculated at CIF, while exports are calculated at FOB.

17. Import-Export Value by Location of China's Foreign Trade Managing Units refers to actual value of imports and exports carried out by corporations which have been registered by the local customs house and are vested with right to run import export business.

18. Disposable Income of Urban Households refers to the actual income at the disposal of members of the households which can be used for final consumption, other non-compulsory expenditure and savings. This equals to total income minus income tax, personal contribution to social security and subsidy for keeping diaries in being a sample household. The following formula is used:

Disposable income = total household income − income tax − personal contribution to social security - subsidy for keeping diaries for a sampled household

19. Consumption Expenditure of Urban Households refers to total expenditure of households for consumption in daily life, including expenditure on the eight categories of food; clothing; household appliances and services; health care and medical services; transport and communications; recreation, education and cultural services; housing; and miscellaneous goods and services.

20. Total Income refers to the sum of income earned from various sources by the rural households and their members during the reference period, and is classified as income from wages and salaries, household operations, properties and transfers.

21. Net Income refers to the total income of rural households from all sources minus all corresponding expenses. The formula for calculation is as follows:

Net income = total income − taxes and fees paid − household operation expenses − taxes and fees − depreciation of fixed assets for production − gifts to non-rural relatives

Net income is mainly used as input for reinvestment in production and as consumption expenditure of the year, and also used for savings and non-compulsory expenses of various forms. "Per capita net income of farmers" is the level of net income averaged by population, reflecting the average income level of rural households in a given area.

22. Total Population refers to the total number of people alive at a certain point of time within a given area.

The annual statistics on total population is taken at midnight, the 31st of December, not including residents in Taiwan province, Hong Kong and Macao and overseas Chinese.

23. Urban Population refers to all people residing in cities and towns.

24. Employed Persons refers to persons aged 16 and over who are engaged in gainful employment and thus receive remuneration payment or earn business income. This indicator reflects the actual utilization of total labour force during a certain period of time and is often used for the research on China’s economic situation and national power.

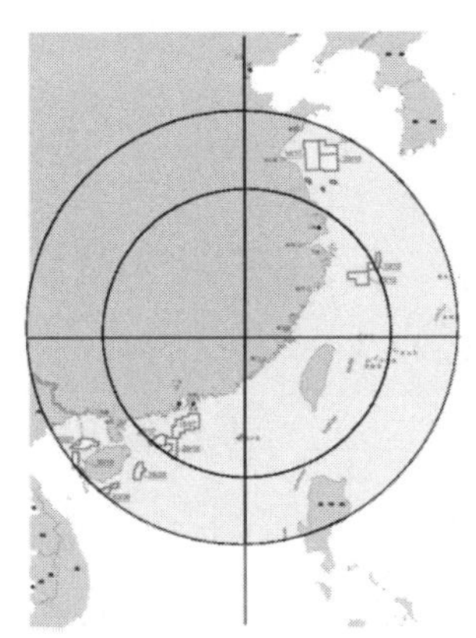

11

部分世界海洋经济统计资料

Part of the World′s Marine Economic Statistics Data

11-1 世界海洋面积
The World Ocean Area

区 域 Region	海洋面积（平方千米） Ocean Area (km^2)	占世界海洋面积的比重（%） Proportion in the World Ocean Area (%)	占地球表面面积的比重（%） Proportion in the Earth's Surface Area (%)
合 计 Total	**361 000 000**	**100.0**	**70.8**
太 平 洋 Pacific Ocean	178 334 000	49.4	35.0
大 西 洋 Atlantic Ocean	91 694 000	25.4	18.0
印 度 洋 Indian Ocean	76 171 000	21.1	14.9
北 冰 洋 Arctic Ocean	14 801 000	4.1	2.9

资料来源：《2011国际统计年鉴》。
Source: *International Statistical Yearbook 2011*.

11-2 主要沿海国家（地区）海岸线长度
Length of Coastline of Major Coastal Countries (Regions)

单位：公里 (km)

国家和地区 Country and Region	海岸线长度 Length of Coastline
美 国 United States	22 680
日 本 Japan	30 000
德 国 Germany	1 300
英 国 United Kingdom	11 450
法 国 France	3 000
意大利 Italy	7 000
加拿大 Canada	20 000
澳大利亚 Australia	20 125
俄罗斯联邦 Russian Fed.	34 000
波 兰 Poland	491
印 度 India	6 083
印度尼西亚 Indonesia	35 000
菲律宾 Philippines	18 533
泰 国 Thailand	3 219
马来西亚 Malaysia	4 675
新加坡 Singapore	193
缅 甸 Myanmar	3 060
孟加拉国 Bangladesh	580
土耳其 Turkey	7 200
韩 国 Republic of Korea	2 413
埃 及 Egypt	2 450
墨西哥 Mexico	9 330
巴 西 Brazil	7 400
阿根廷 Argentina	4 989

11-3 世界主要沿海国家（地区）土地和人口（2009年）
Land Area and Population of Major Coastal Countries (Regions), 2009

国家和地区 Country and Region	国土面积（万平方千米） Area of Territory (10 000 km^2)	年中人口（万人） Mid-year Population (10 000 persons)	人口密度（人/平方千米） Population Density (people per km^2)
世界总计 World Total	**13 412.3**	**677 524**	**52**
中 国 China	960.0	133 146	143
美 国 United States	983.2	30 701	34
日 本 Japan	37.8	12 756	350
加拿大 Canada	998.5	3 374	4
德 国 Germany	35.7	8 188	235
英 国 United Kingdom	24.4	6 184	256
法 国 France	54.9	6 262	114
意大利 Italy	30.1	6 022	205
墨西哥 Mexico	196.4	10 743	55
印度尼西亚 Indonesia	190.5	22 996	127
马来西亚 Malaysia	33.1	2 747	84
泰 国 Thailand	51.3	6 776	133
新加坡 Singapore	0.1	499	7 125
韩 国 Republic of Korea	10.0	4 875	503
印 度 India	328.7	115 535	389
巴 西 Brazil	851.5	19 373	23
俄罗斯联邦 Russian Fed.	1 709.8	14 185	9

注：数据来源于《2011年中国统计年鉴》。

Note: The data come from the *China Statistical Yearbook 2011*.

11-4 主要沿海国家（地区）国内生产总值
Gross Domestic Product of Major Coastal Countries (Regions)

国家和地区 Country and Region	2009年			2010年
	国内生产总值（亿美元）GDP (100 million US$)	人均国民总收入（美元）GNI per Captia (current US$)	国内生产总值增长率（%）Growth Rate of GDP(%)	国内生产总值①（亿美元）GDP (100 million US$)
美国 United States	142 563	47 240	-2.4	146 578
日本 Japan	50 675	37 870	-5.2	54 589
德国 Germany	33 467	42 560	-5.0	33 156
英国 United Kingdom	21 745	41 520	-4.9	22 475
法国 France	26 494 ②	42 680 ②	-2.2	25 825
意大利 Italy	21 128	35 080	-5.0	20 551
加拿大 Canada	13 361	42 170	-2.6	15 741
俄罗斯联邦 Russian Fed.	12 307	9 370	-7.9	14 651
印度 India	13 102	1 180	5.7	15 380
印度尼西亚 Indonesia	5 403	2 230	4.6	7 067
泰国 Thailand	2 639	3 760	-2.3	3 189
马来西亚 Malaysia	1 916	7 230	-1.7	2 380
新加坡 Singapore	1 822	37 220	-2.0	2 227
韩国 Republic of Korea	8 325	19 830	0.2	10 071
墨西哥 Mexico	8 749	8 920	6.5	10 391
巴西 Brazil	15 720	8 040	-0.2	20 903

注：①按汇率法计算。

②包括法属圭亚那、瓜德罗普、马提尼克和留尼汪。

Note: ①Estimated by market exchnge rates.

②Including French Guiana, Guadeloupe, Martinique and Reunion.

11-5 世界主要沿海国家(地区)国内生产总值产业构成(2009年)
Industrial Composition of GDP of Major Coastal Countries (Regions), 2009

国家和地区 Country and Region	国内生产总值产业构成（%） Industry Structure of GDP(%)		
	第一产业 Primary Industy	第二产业 Secondary Industry	第三产业 Tertiary Industy
世　界 World	**2.9** ②	**27.5** ②	**69.4** ②
美　国 United States	1.3 ②	21.4 ②	77.4 ②
日　本 Japan	1.5 ②	29.3 ②	69.3 ②
德　国 Germany	0.9 ①	30.1 ①	69.0 ①
英　国 United Kingdom	0.7 ①	23.6 ①	75.7 ①
法　国 France	2.0 ①	20.5 ①	77.5 ①
意大利 Italy	2.0 ①	27.1 ①	70.9 ①
墨西哥 Mexico	4.2	37.6	58.3
印度尼西亚 Indonesia	14.1	47.0	39.2
马来西亚 Malaysia	8.7	55.4	35.9
泰　国 Thailand	11.7 ①	44.2 ①	44.2 ①
新加坡 Singapore	0.1 ①	26.3	74.0 ①
韩　国 Republic of Korea	2.7 ①	36.5 ①	60.9 ①
印　度 India	17.1	28.3	54.6
巴　西 Brazil	6.6	27.2	66.2
俄罗斯联邦 Russian Fed.	5.0 ①	37.2 ①	57.8 ①

注：①2008年数据。②2007年数据。

Note: ① The data for 2008.② The data for 2007.

11-6 世界主要沿海国家（地区）就业人数
Employment in the Major Coastal Countries (Regions)

单位：万人　　(10 000 persons)

国家和地区 Country and Region	2000	2004	2005	2006	2007	2008
中　　国① China	72 085	75 200	75 825	76 400	76 990	77 480
中国香港② Hong Kong,China	321	327	334	340	348	352
孟加拉国 Bangladesh	5 176		4 736			
印　　度 India	36 897					
印度尼西亚 Indonesia	8 984	9 372	9 396	9 546	9 993	10 255
以 色 列④ Israel	222	240	249	257	268	278
日　　本 Japan	6 446	6 329	6 356	6 382	6 412	6 385
哈萨克斯坦 Kazakhstan	620	718	726	740	763	786
韩　　国④ Republic of Korea	2 116	2 256	2 286	2 315	2 343	2 358
马来西亚④⑤ Malaysia	927	998	1 005	1 028	1 054	1 066
巴基斯坦③④ Pakistan	3 685	4 201	4 292	4 695	4 765	
菲 律 宾④ Philippines	2 745	3 161	3 231	3 264	3 356	3 409
新 加 坡 Singapore	148	163		180	180	185
斯里兰卡⑦ Sri Lanka	631③	739⑧	752	711	704	717
泰　　国④ Thailand	3 300⑨	3 571	3 630	3 634	3 712	3 784
土 耳 其④ Turkey	2 158	2 179	2 205	2 233	2 074	2 119
越　　南 Viet Nam	3 837	4 232	4 253	4 334	4 417	4 492
埃　　及④⑤ Egypt	1 720	1 872	1 934	2 044	2 172	

11-6 续表 continued

国家和地区 Country and Region	2000	2004	2005	2006	2007	2008
南　　非 South Africa	1 224	1 164	1 230	1 280	1 323	1 371
加拿大④⑩ Canada	1 476	1 595	1 617	1 648	1 687	1 713
墨西哥⑪ Mexico	3 804	4 056	4 079	4 220	4 291	4 387
美　　国④⑥ United States	13 521	13 925	14 173	14 443	14 605	14 536
阿根廷③⑫ Argentina	826	942	964	1 004	1 012	1 028
巴　　西③ Brazil	6 563	8 460	8 719	8 932	9 079	
委内瑞拉④ Venezuela	896③	1042③	1073③	1112③	1 149	1 186
法　　国 France	2 326	2 480	2 498	2 513	2 557	2 591
德　　国 Germany	3 660	3 566	3 657	3 732	3 816	3 873
意大利 Italy	2 123	2 240	2 256	2 299	2 322	2 340
荷　　兰 Netherlands	780	793	796	811	831	846
波　　兰④ Poland	1 453	1 380	1 412	1 459	1 524	1 580
俄罗斯联邦⑬ Russian Fed.	6 507	6 728	6 817	6 886	7 057	7 097
西班牙④⑥ Spain	1 551	1 797	1 897	1 975	2 036	2 026
乌克兰⑭ Ukraine	2 018	2 030	2 068	2 073	2 090	2 097
英　　国⑥ United Kingdom	2 740	2 837	2 867	2 893	2 910	2 948
澳大利亚④ Australia	895	962	997	1 022	1 051	1 074
新西兰④ New Zealand	180	202	208	213	217	219

注：①不包括军人和返聘的退休人员。②不包括军人和域外机构人员。③10岁及以上。④不包括军人。⑤15岁至64岁。⑥16岁及以上。⑦不包括北部和东部省。⑧不包括北部省。⑨13岁及以上。⑩不包括生活在保护的居民。⑪14岁以上。⑫部分城市地区 ⑬15岁至72岁。⑭15岁至70岁。

Note: ①Excluding armed forces and reemployed retired persons. ②Excluding marine, military and institutional populations. ③Persons aged 10 years. ④Excluding armed forces. ⑤Persons aged 15 to 64 years. ⑥Persons aged 16 years and over. ⑦Excluding Northern and Eastern provinces. ⑧Excluding Northern provinces. ⑨Persons aged 13 years and over.⑩Excluding residents of the territories and indigenous persons living on reserves.⑪Person aged 14 years and over.⑫Some urban agglomerations.⑬Persons aged 15 to 72 years.⑭Persons aged 15 to 70 years.

11-7 主要沿海国家（地区）渔业增加值
Fishery Added Value of Major Coastal Countries (Regions)

单位：亿本币 (100 million local currency units)

国家和地区 Country and Region	渔业增加值 Fishery Added Value					
	2000	2004	2005	2006	2007	2008
中国 China	14 900	21 413	22 400	24 000	28 600 ①	33 700 ①
中国香港 Hong Kong, China	9	9	8	8	9 ①	8 ①
孟加拉国 Bangladesh	1 370	1 478	1 550	1 630	1 780	1 980
印度 India	2 150	2 796	3 110	3 560	3 910	4 270
印度尼西亚② Indonesia	30	53	60	74	98	137
伊朗② Iran	2	3	4	5	5	
日本② Japan	1	8				
哈萨克斯坦 Kazakhstan	40	58	60	73	103	125
韩国 Republic of Korea	21 547	18 110	17 226	15 599	14 832	20 000 ②
马来西亚 Malaysia	51	53	56	66	68	75
菲律宾 Philippines	785	1 105	1 160	1 300	1 430	1 700
新加坡 Singapore	2 ①	2 ①	2 ①	2 ①	2 ①	2 ①
泰国 Thailand	1 180	1 070	1 040	1 090	1 030	1 060
越南② Viet Nam	15	27	33	38	46	58
加拿大 Canada	14	14	14	13		
墨西哥 Mexico	64	84	73	62		
美国 United States	980	1 422	1 288	1 254		
阿根廷 Argentina	6	13	16	20	18	24
委内瑞拉 Venezuela	2 200	1 ②	9 500	9 880		
法国 France	14	15	15	14	14	13
德国 Germany	2	2	2	2	2	2
意大利 Italy	13	15	15	16		
荷兰 Netherlands	3	2	2	2	2	2
波兰 Poland	2	2	2	2	2	1
西班牙 Spain	15	17	16	16	17	17
乌克兰 Ukraine	2	1	2	2	2	2
英国 United Kingdom	4	4	4			
澳大利亚 Australia	249	268	275	223		
新西兰 New Zealand	4			2		

注：①为农业、狩猎业、林业和渔业增加值。②万亿本币。

Note:① The added value of agriculture, hunting, forestry and fishery.
②Trillion local currency units.

11-8 主要沿海国家（地区）旅馆和饭店业增加值
Added Value of Hotel Industry in the Major Coastal Countries (Regions)

单位：亿本币 (100 million local currency units)

国家和地区 Country and Region	旅馆和饭店业增加值 Added Value of Hotel Industry					
	2000	2004	2005	2006	2007	2008
中国 China	2 150	3 665	4 200	4 790	5 550	6 620
中国香港 Hong Kong, China	366	331	365	414	476	
孟加拉国 Bangladesh	146	220	251	285	329	389
印度 India	2 510	4 127	5 440	6 610	7 790	8 140
伊朗* Iran	2	6	8	9	11	
哈萨克斯坦 Kazakhstan	148	516	681	842	1 140	1 320
韩国 Republic of Korea	148 618	184 206	188 713	197 262	205 598	23 *
马来西亚 Malaysia	80	102	112	124	142	163
菲律宾 Philippines	630	894	959	1 050	1 190	1 300
新加坡 Singapore	34	33	38	43	50	55
泰国 Thailand	2 750	3 342	3 470	3 860	4 170	4 380
越南* Viet Nam	14	23	29	36	45	65
埃及 Egypt	66	127	177	247		
加拿大 Canada	234	271	283	296		
墨西哥 Mexico	2 470	3 067	2 380	2 500	2 650	
美国 United States	2 610	3 137	3 340	3 580	3 800	
阿根廷 Argentina	78	99	126	163	200	251
委内瑞拉 Venezuela	14 000	3 *	55 500	80 800		
法国 France	302	346	367	383	405	408
德国 Germany	301	318	331	337	370	380
意大利 Italy	416	464	484	504		
荷兰 Netherlands	76	83	85	88	93	90
波兰 Poland	83	95	107	111	119	138
西班牙 Spain	434	574	610	652	681	722
乌克兰 Ukraine	8	23	24	54	68	97
英国 United Kingdom	260	331	327			
澳大利亚 Australia	183	194	238	256	269	285
新西兰 New Zealand	21			31		

注：*万亿本币。

Note: *Trillion local currency units.

11-9 主要沿海国家（地区）运输、仓储和通讯业增加值
Added Values of Transportation,Storage and Communications Industries in the Major Coastal Countries (Regions)

单位：亿本币 (100 million local currency units)

国家和地区 Country and Region	运输、仓储和通讯业增加值 Added Values of Transportation, Storage and Commnuications Industries					
	2000	2004	2005	2006	2007	2008
中国 China	6 160	9 304	10 700	12 200	14 600	16 400
中国香港 Hong Kong,China	1 190	1 268	1 350	1 370	1 420	1 220
孟加拉国 Bangladesh	1 970	3 444	3 830	4 320	4 890	5 690
印度 India	15 000	24 801	28 300	32 500	36 700	42 000
印度尼西亚* Indonesia	65	142	181	232	264	312
伊朗* Iran	48	98	148	191	240	
以色列 Israel	342	398	417	406	450	465
日本* Japan	35	34	34	34	34	
哈萨克斯坦 Kazakhstan	2 990	6 912	8 970	11 800	14 800	17 700
韩 国 Republic of Korea	361 387	509 690	524 295	538 143	574 511	610 000
马来西亚 Malaysia	249	333	360	389	426	458
巴基斯坦 Pakistan	4 010	6 756	7 600	9 080	10 100	11 800
菲律宾 Philippines	1 990	3 674	4 140	4 460	4 780	5 090
新加坡 Singapore	208	256	280	289	334	343
斯里兰卡 Sri Lanka	1 350	2 426	2 910	3 490	4 280	5 370
泰国 Thailand	3 960	4 925	5 200	5 690	6 260	6 430
越南* Viet Nam	17	30	37	44	51	67
埃及 Egypt	257	468	574	580	767	924
南非 South Africa	809	1 222	1 390	1 540	1 640	1 880
加拿大 Canada	702	838	918	981		
墨西哥 Mexico	5 570	7 255	8 230	9 240	10 100	
美国 United States	6 360	7 136	7 390	7 720	8 190	
阿根廷 Argentina	241	374	444	536	641	803
巴西 Brazil	866	1 425	1 650	1 760	1 980	
委内瑞拉 Venezuela	52 900	12 *	164 000	237 000		
法国 France	776	965	998	1 030	1 080	1 120
德国 Germany	1 020	1 174	1 160	1 220	1 230	1 280
意大利 Italy	777	960	981	1 009		
荷兰 Netherlands	266	323	332	339	351	350
波兰 Poland	434	608	627	685	722	783
俄罗斯联邦 Russian Fed.	5 930	16 620	19 300	22 800	27 500	32 100
西班牙 Spain	418	546	563	601	641	663
乌克兰 Ukraine	198	427	474	561	701	871
英 国 United Kingdom	693	793	811			
澳大利亚 Australia	590	652	804	899	949	1 060
新西兰 New Zealand	82			117		

注：*万亿本币。
Note: *Trillion local currency units.

11-10 主要沿海国家（地区）海洋渔区标称渔获量
Nominal Catches of Marine Fishing Zone in the Major Coastal Countries (Regions)

国家和地区 Country and Region	2007	2008
世界总计 World Total	**89 898 882**	**89 740 919**
阿尔巴尼亚 Albania	2 899	
阿尔及利亚 Algeria	148 436	
安哥拉 Angola	312 436	317 262
安提瓜和巴布达 Antigua Barb	3 092	
阿根廷 Argentina	985 409	995 083
澳大利亚 Australia	183 756	
巴哈马 Bahamas	3 749	
巴林 Bahrain	15 012	
孟加拉国 Bangladesh	1 494 199	1 557 754
巴巴多斯 Barbados	1 800 *	
比利时 Belgium	24 029	
伯利兹 Belize	6 682	
贝宁 Benin	7 701	
巴西 Brazil	783 177	775 000 *
文莱 Brunei Darsm	2 241	
保加利亚 Bulgaria	7 829	
柬埔寨 Cambodia	457 000 *	431 000
喀麦隆 Cameroon	64 232	
加拿大 Canada	1 005 955	937 370
佛得角群岛 Cape Verde	18 328	
智利 Chile	3 819 303	3 554 814
中国台湾 China, Taiwan	1 174 393	1 016 390
哥伦比亚 Colombia	76 000 *	
科摩罗群岛 Comoros	16 000	
刚果 Congo Rep.	29 821	
库克群岛 Cook Is.	3 200	
哥斯达黎加 Costa Rica	20 735	
科特迪瓦 Côte dIvoire	26 100 *	
古巴 Cuba	33 572	
塞浦路斯 Cyprus	2 426	
丹麦 Denmark	652 919	690 202
多米尼加共和国 Dominican Rp.	12 228	
厄瓜多尔 Ecuador	384 267	434 239
埃及 Egypt	372 491	373 815
萨尔瓦多 El Salvador	46 217	

11-10 续表1 continued

国家和地区 Country and Region	2007	2008
赤道几内亚 Eq. Guinea	2 883	
爱沙尼亚 Estonia	95 171	
法罗群岛 Faeroe Is.	582 134	495 348
福克兰群岛 Falkland Is.	72 146	
斐济 Fiji	46 489	
芬兰 Finland	128 169	
法国 France	513 392	457 127 *
波利尼西亚 Fr. Polynesia	13 027	
加蓬 Gabon	29 500 *	
冈比亚 Gambia	38 709	
格鲁吉亚 Georgia	18 147	
德国 Germany	248 764	229 499
加纳 Ghana	320 730	349 831
希腊 Greece	94 655	
格陵兰岛 Greenland	233 754 *	233 754 *
格林纳达 Grenada	2 407	
瓜德罗普 Guadeloupe	10 100	
危地马拉 Guatemala	15 227	
几内亚 Guinea	96 000 *	
几内亚比绍 Guinea Bissau	6 050 *	
圭亚那 Guyana	46 640	
海地 Haiti	9 700 *	
洪都拉斯 Honduras	12 778 *	
冰岛 Iceland	1 399 167	1 284 034
印度 India	3 859 293	4 104 877
印度尼西亚 Indonesia	5 050 340	4 957 098
伊朗 Iran	403 635	407 842
伊拉克 Iraq	12 319	
爱尔兰 Ireland	214 871	205 342
以色列 Israel	2 220 *	
意大利 Italy	286 645	235 785
牙买加 Jamaica	16 148	
日本 Japan	4 296 532	4 248 697
肯尼亚 Kenya	7 448	
基里巴斯 Kiribati	21 598 *	
朝鲜 Korea D. P. Rp.	205 000 *	205 000 *
韩国 Korea Rep.	1 869 840	1 943 870

11-10 续表2 continued

国家和地区 Country and Region	2007	2008
拉脱维亚 Latvia	154 966	
黎巴嫩 Lebanon	3 541	
利比里亚 Liberia	12 745	
利比亚 Libya	31 921	
立陶宛 Lithuania	185 639	
马达加斯加 Madagascar	115 148	
马来西亚 Malaysia	1 385 703	1 395 942
马尔代夫 Maldives	143 597	
马耳他 Malta	1 235	
马尔提尼克 Martinique	6 300	
毛里塔尼亚 Mauritania	186 588 *	
毛里求斯 Mauritius	7 906	
墨西哥 Mexico	1 483 749 *	1 588 857 *
密克罗尼西亚 Micronesia	16 985	
摩洛哥 Morocco	879 273 *	995 773
莫桑比克 Mozambique	68 189	
缅甸 Myanmar	2 235 580	2 493 750
纳米比亚 Namibia	413 333	372 822
荷兰 Netherlands	413 602	416 748
荷属安的列斯群岛 Neth Antilles	3 662	
新喀里多尼亚 New Caledonia	3 510	
新西兰 New Zealand	494 492	451 052
尼加拉瓜 Nicaragua	24 598	
尼日利亚 Nigeria	530 420	541 368
挪威 Norway	2 378 841	2 430 842
阿曼 Oman	151 744	
巴基斯坦 Pakistan	440 188	451 414
帛琉群岛 Palau Is.	985	
巴拿马 Panama	208 528	222 508
巴布亚新几内亚 Papua New Guinea	268 874	223 631
秘鲁 Peru	7 210 544	7 362 907
菲律宾 Philippines	2 499 680	2 561 192
波兰 Poland	132 761	
葡萄牙 Portugal	238 367	240 192
波多黎各 Puerto Rico	1 675	
卡塔尔 Qatar	15 190	
留尼汪岛 Reunion	3 924	

11-10 续表3 continued

国家和地区 Country and Region	2007	2008
俄罗斯 Russian Fed.	3 454 218	3 383 724
萨摩亚群岛 Samoa Is.	4 605	
圣多美和普林西比 Sao Tome & Principe	4 150 *	
沙特阿拉伯 Saudi Arabia	70 000 *	
塞内加尔 Senegal	435 502	447 754
塞舌尔 Seychelles	65 871	
塞拉利昂 Sierra Leone	144 535	203 582
新加坡 Singapore	3 522	
索罗门群岛 Solomon Is.	31 272 *	
索马里 Somalia	29 800 *	
南非 South Africa	678 757	643 686
西班牙 Spain	820 118	917 188
斯里兰卡 Sri Lanka	307 279	327 575
苏丹 Sudan	5 699	
苏里南 Suriname	29 277	
瑞典 Sweden	238 253	231 336
叙利亚 Syria	3 381	
坦桑尼亚 Tanzania	328 527	325 476
泰国 Thailand	2 304 957	2 457 184
多哥 Togo	14 905	
汤加 Tonga	2 545 *	
特立尼达和多巴哥 Trinidad & Tobago	8 406	
突尼斯 Tunisia	102 079	
土耳其 Turkey	632 450	494 124
特克斯和凯科斯群岛 Turks & Caicos	4 830	
乌克兰 Ukraine	200 905	
阿拉伯联合酋长国 United Arab Em.	87 000 *	
英国 United Kingdom	619 691	596 004
美国 United States	4 767 596	4 349 853
乌拉圭 Uruguay	107 520	
瓦努阿图 Vanuatu	85 356	
委内瑞拉 Venezuela	256 353	295 364
越南 Viet Nam	2 020 400	2 087 500
也门 Yemen	179 916	

注:*为联合国粮农组织估算值 (表11-11同)。
资料来源：《渔业统计年鉴》，联合国粮农组织，2008年 (表11-11同) 。

Note: * It is estimated by FAO from available sources of information or calculation.(Same as Table 11-11).
Source: *Yearbook of Fishery*, FAO, 2008 (Same as Table 11-11).

11-11 主要沿海国家（地区）海藻和其他养殖品种捕捞产量
Capture Production of Seaweeds and Other Aquatic Plants in the Major Coastal Countries (Regions)

单位：吨 (t)

国家和地区 Country and Region	2006	2007	2008
世界总计 World Total	**1 011 887**	**1 076 331**	**1 045 069**
阿根廷 Argentina	...	...	...
澳大利亚 Australia	15 504	2 223	1 923
加拿大 Canada	11 313	19 385	12 614
智利 Chile	301 115	312 160	384 563
爱沙尼亚 Estonia	394	1 608	1 483
斐济 Fiji	250 *	200 *	150
法国 France	19 160	39 757	39 757 *
冰岛 Iceland	20 964	21 867	22 559
印度尼西亚 Indonesia	4 996	4 643	2 917
意大利 Italy	1 400 *	1 400 *	1 400 *
日本 Japan	113 665	103 602	104 700 *
韩国 Korea Rep.	13 754	18 189	13 866
马达加斯加 Madagascar	5 300	3 650	3 650
墨西哥 Mexico	4 532	5 093	4 900
摩洛哥 Morocco	14 870	12 373	9 037
纳米比亚 Namibia	...	...	...
挪威 Norway	145 429	134 671	...
秘鲁 Peru	3 434	10 786	13 779
菲律宾 Philippines	314	351	388
葡萄牙 Portugal	765	495	198
俄罗斯 Russian Fed.	11 614	8 342	10 242
塞内加尔 Senegal			
南非 South Africa	9 776 *	12 472 *	11 767
西班牙 Spain	485	109	97
坦桑尼亚 Tanzania	278	214	277
乌克兰 Ukraine	1 121	1 892	1 947
美国 United States	6 362	2 272	6 951

11-12 主要沿海国家（地区）油气可开采储量（2007年）
Recoverable Reserves of Oil and Gas in the Major Coastal Countries (Regions), 2007

国家和地区 Country and Region	原油和液化天然气（万吨） Crude Oil and liquified Natural Gas (10 000 t)	天然气（亿立方米） Natural Gas (100 million m^3)	油页岩(万吨) Oil Shale (10 000 t)
中国 China	30 000	246 600	229 000
孟加拉国 Bangladesh	3 920	300	
印度 India	10 550	72 500	
印度尼西亚 Indonesia	27 540	52 900	
伊朗 Iran	280 800	1 845 000	
以色列 Israel	340		55 000
日本 Japan	510	600	
马来西亚 Malaysia	24 800	48 700	
缅甸 Myanmar	5 150	700	28 600
巴基斯坦 Pakistan	8 440	4 000	
菲律宾 Philippines	950	1 600	
泰国 Thailand	3 170	4 900	91 600
土耳其 Turkey	150	16 500	28 400
越南 Viet Nam	2 200	45 300	
埃及 Egypt	21 520	56 100	81 600
尼日利亚 Nigeria	53 080	486 000	
南非① South Africa	100	200	1 900
加拿大 Canada	16 340	281 800	219 200
墨西哥 Mexico	3 730	164 500	
美国 United States	68 230	371 700	30 156 600
阿根廷 Argentina	4 420	35 800	5 700
巴西 Brazil	3 650	170 600	1 173 400
委内瑞拉 Venezuela	48 380	1 399 700	
法国② France	100	1 600	100 200
德国 Germany	2 180	3 700	28 600
意大利③ Italy	1 700	10 600	1 044 600
荷兰 Netherlands	12 710	1 200	
波兰 Poland	750	1 600	700
俄罗斯联邦 Russian Fed.	478 140	1 070 000	3 547 000
西班牙 Spain	30	2 000	4 000
英国 United Kingdom	3 430	45 200	50 100
澳大利亚 Australia	9 070	26 400	453 100
新西兰 New Zealand	560	1 800	300

注：①包括博茨瓦纳、莱索托、斯威士兰和纳米比亚。②包括摩纳哥。③包括圣马力诺。
Note: ①Includes Botswana, Lesotho, Swaziland and Namibia. ②Includes Monaco.③Includes San Marino.

11-13 主要沿海国家（地区）油气产量（2009年）
Natural Gas and Crude Oil Production of the Major Coastal Countries (Regions) 2009

国家和地区 Country and Region	天然气产量（万亿焦耳） Production of Natural Gas (tera Joule)	原油产量（万吨） Production of Crude Oil (10 000 t)
中国 China	3 325 488	18 960
孟加拉国 Bangladesh	687 048	
印度 India	1 570 416	3 323
印度尼西亚 Indonesia	3 309 552	4 657
哈萨克斯坦 Kazakhstan	381 696	7 658
马来西亚 Malaysia	2 279 292	3 143
泰国 Thailand	940 308	1 180
加拿大 Canada	5 685 504	12 564
墨西哥 Mexico	3 235 536	13 660
美国 United States	22 763 412	35 791
阿根廷 Argentina	1 825 728	3 584
巴西 Brazil	924 732	9 914
德国 Germany	591 216	438
意大利 Italy	309 336	444
荷兰 Netherlands	2 341 464	
波兰 Poland	169 932	
俄罗斯联邦 Russian Fed.	19 656 516	49 404
乌克兰 Ukraine	712 944	391
英国 United Kingdom	2 496 888	6 235
澳大利亚 Australia	1 506 372	2 369
新西兰 New Zealand		251

11-14 主要沿海国家（地区）能源利用效率
Use Efficiency of Energy, Resources in the Major Coastal Countries (Regions)

单位：吨标准油/万美元 (ton of oil equivalent per 10 000 US$)

国家和地区 Country and Region	单位国内生产总值能耗 Energy consumption per unit of GDP		
	2005	2006	2007
世界 World	**3.02**	**2.98**	**2.94**
中国 China	8.85	8.58	7.96
中国香港 Hong Kong,China	0.63	0.60	0.58
孟加拉国 Bangladesh	3.89	3.77	3.70
印 度 India	8.29	7.95	7.69
印度尼西亚 Indonesia	8.43	8.23	8.17
伊朗 Iran	11.93	12.23	12.18
以色列 Israel	1.47	1.45	1.43
日 本 Japan	1.04	1.02	0.99
哈萨克斯坦 Kazakhstan	18.75	19.14	18.40
韩 国 Republic of Korea	3.17	3.06	3.03
马来西亚 Malaysia	5.52	5.33	5.46
巴基斯坦 Pakistan	7.99	7.84	7.86
菲律宾 Philippines	4.20	3.96	3.75
新加坡 Singapore	2.29	2.06	1.87
斯里兰卡 Sri Lanka	4.54	4.25	4.07
泰 国 Thailand	6.18	6.03	5.99
土耳其 Turkey	5.53	2.61	2.68
越南 Viet Nam	11.39	10.88	10.62
埃 及 Egypt	5.12	5.03	4.95
尼日利亚 Nigeria	16.93	15.95	15.24
南非 South Africa	7.90	7.63	7.52
加拿大 Canada	3.23	3.12	3.05
墨西哥 Mexico	2.75	2.62	2.67
美 国 United States	2.12	2.05	2.04
阿根廷 Argentina	2.01	2.01	1.98
巴 西 Brazil	2.92	2.90	2.89
委内瑞拉 Venezuela	4.48	4.18	4.04
法 国 France	1.88	1.82	1.75
德 国 Germany	1.73	1.69	1.60
意大利 Italy	1.59	1.55	1.50
荷兰 Netherlands	1.92	1.80	1.83
波 兰 Poland	4.63	4.59	4.29
俄罗斯联邦 Russian Fed.	18.62	17.80	16.50
西班牙 Spain	2.08	1.99	1.96
乌克兰 Ukraine	31.59	28.30	26.23
英 国 United Kingdom	1.33	1.28	1.20
澳大利亚 Australia	2.48	2.44	2.38
新西兰 New Zealand	2.65	2.62	2.57

11-15 主要沿海国家（地区）国际海运装货量和卸货量
International Ocean Shipping Loading and Unloading Capacities in the Major Coastal Countries (Regions)

单位：万吨 (10 000 t)

国家和地区 Country and Region	国际海运装货量 International Ocean Shipping Loading Capacity		国际海运卸货量 International Ocean Shipping Unloading Capacity	
	2007	2008	2007	2008
中国香港 Hong Kong,China	10 411	11 342	14 130	14 598
孟加拉国 Bangladesh	73	92	1 810	1 801
印度尼西亚 Indonesia	34 850	36 390	9 172	9 994
伊朗 Iran	3 943	3 058	7 790	6 611
以色列 Israel	1 775	1 800	2 248	2 254
韩国 Republic of Korea	28 602		57 650	
马来西亚 Malaysia	9 697	10 367	11 708	12 588
新加坡 Singapore	48 361 ①	51 541 ①		
阿根廷② Argentina	2 299		2 299	2 640
德国 Germany	11 880	11 990	18 810	19 250
波兰 Poland	2 632	2 076	2 612	2 807
乌克兰 Ukraine	6 386	7 626	1 834	2 114
澳大利亚 Australia	67 078	73 348	7 993	8 718
新西兰 New Zealand	2 341	2 525	1 847	1 940

注：①包括装货量和卸货量。②不包括转口。

Note: ①Including both goods loaded and unloaded.②Excluding re-exports.

11-16 世界主要外贸海运量（2009年）
World Major Maritime Freight Traffic in Foreign Trade, 2009

品 种 Sort	海运量（百万吨） Freight Traffic (million tons)		海运周转量（十亿吨海里） Ton-Kilometer (billion ton-sea mile)	
	2009*	所占比例（%） Percentage	2008*	所占比例（%） Percentage
合 计 **Total**	**7 850**	**100.0**	**52 515**	**100.0**
原 油 Crude Oil	1 920	24.5	14 967	28.5
成品油 Refined Oil	746	9.5	3 206	6.1
铁矿石 Ironstone	849	10.8	8 361	15.9
煤 炭 Coal	777	9.9	6 284	12.0
谷 物 Corn	320	4.1	3 153	6.0
其 他 Others	3 238	41.2	16 544	31.5

注：*为估计数。
资料来源:《海运统计与市场评论》1月/2月 2010。
Note: *Estimated data.
Source: *Shipping Statistics and Market Review*, January/February2010, ISL.

11-17 主要沿海国家（地区）国际旅游人数
Number of International Tourists of Major Coastal Countries (Regions)

单位：万人 (10 000 persons)

国家和地区 Country and Region	国外游客到达人数 Arrivies Number of Foreign Tourists			出国旅游人数 Number of Tourists Going Abroad		
	2000	2007	2008	2000	2007	2008
世界总计 World Total	**68 922**	**91 202**	**92 785**	**75 564**	**100 179**	**102 706**
高收入国家 High Income	45 849	54 744	54 653	41 676	52 814	53 145
中等收入国家 Middle Income	21 026	33 130	34 587	24 700	36 313	
低收入国家 Low Income	1 103	2 374				
中国 China	3 123	5 472	5 305	1 047	4 095	4 584
孟加拉国 Bangladesh	20	29	47	113	233	88
印 度 India	265	508	537	442	978	1 065
印度尼西亚 Indonesia	506	551	623		516	549
伊朗 Iran	134	222	203	229		
以色列 Israel	242	207	257	353	415	421
日 本 Japan	476	835	835	1 782	1 730	1 599
哈萨克斯坦 Kazakhstan	147	388	345	125	454	524
韩 国 Republic of Korea	532	645	689	551	1 333	1 200
马来西亚 Malaysia	1 022	2 097	2 205	3 053		
缅 甸 Myanmar	21	25	19			
巴基斯坦 Pakistan	56	84	82			
菲律宾 Philippines	199	309	314	167		
新加坡 Singapore	606	796	778	444	602	683

11-17 续表 continued

国家和地区 Country and Region	国外游客到达人数 Arrivies Number of Foreign Tourists			出国旅游人数 Number of Tourists Going Abroad		
	2000	2007	2008	2000	2007	2008
斯里兰卡 Sri Lanka	40	49	44	52	86	97
泰 国 Thailand	958	1 446	1 454	191	402	
土耳其 Turkey	959	2 225	2 499	528	894	987
埃 及 Egypt	512	1 061	1 230	296		
尼日利亚 Nigeria	81					
南非 South Africa	587	909	959	383	443	443
加拿大 Canada	1 963	1 794	1 714	1 918	2 516	2 704
墨西哥 Mexico	2 064	2 137	2 264	1 108	1 508	1 445
美 国 United States	5 124	5 598	5 794	6 133	6 402	6 355
阿根廷 Argentina	291	456	467	495	417	461
巴 西 Brazil	531	503	505	323	482	494
委内瑞拉 Venezuela	47	77	75	95	141	175
法 国 France	7 719	8 084	7 845	1 989	2 514	2 335
德 国 Germany	1 898	2 442	2 488	7 440	7 040	7 300
意大利 Italy	4 118	4 365	4 273	2 199	2 773	2 828
荷兰 Netherlands	1 000	1 101	1 010	1 390	1 756	1 846
波 兰 Poland	1 740	1 498	1 296	5 668	4 756	
俄罗斯联邦 Russian Fed.	2 117	2 291	2 368	1 837	3 429	3 654
西班牙 Spain	4 640	5 897	5 732	410	1 128	1 123
乌克兰 Ukraine	643	2 312	2 545	1 342	1 734	1 550
英 国 United Kingdom	2 321	3 087	3 014	5 684	6 945	6 901
澳大利亚 Australia	493	564	559	350	546	581
新西兰 New Zealand	178	243	241	128	198	197

11-18 集装箱吞吐量居世界前20位的港口
World Top Twenty Seaports in Terms to the Number of Containers Handled

单位：万标准箱 (10 000 TEU)

港　口 Seaport	所属国家或地区 Country or Region	吞吐量 Containers Handled
上海 Shanghai	中国 China	2 907
新加坡 Singapore	新加坡 Singapore	2 843
香港 Hong Kong	中国 China	2 370
深圳 Shenzhen	中国 China	2 251
釜山 Pusan	韩国 Republic of Korea	1 416
宁波 舟山 Ningbo and Zhoushan	中国 China	1 315
广州 Guangzhou	中国 China	1 255
青岛 Qingdao	中国 China	1 201
迪拜 Dubayy	阿联酋 United Arab Em	1 160
鹿特丹 Rotterdam	荷兰 Netherlands	1 110
天津 Tianjin	中国 China	1 009
高雄 Gaoxiong	中国台湾 Taiwan, China	918
巴生港 Kelang	马来西亚 Malaysia	840
安特卫普 Antwerp	比利时 Belgium	848
汉堡 Hamburg	德国 Germany	790
洛杉矶 Los Angeles	美国 United States	783
丹戎帕拉帕斯 Tanjung Periuk	马来西亚 Malaysia	653
长滩 Long Beach	美国 United States	626
厦门 Xiamen	中国 China	582
大连 Dalian	中国 China	526

资料来源：上海航运交易所，CL-ONLINE。

Source: Shanghai Shipping Exchange, CL-ONLINE.

11-19 货物吞吐量居世界前20位的港口（2009年）
World Top Twenty Seaports in Terms of the Cargo Handled, 2009

单位：百万吨 (million tons)

港口 Seaport	所属国家或地区 Country or Region	吞吐量 Cargo Handled	备注 Remark
上海 Shanghai	中国 China	592.1	内外贸货物 domestic and foreign trade cargo
新加坡 Singapore	新加坡 Singapore	472.3	内外贸货物 domestic and foreign trade cargo
鹿特丹 Rotterdam	荷兰 Netherlands	387.0	外贸货物 foreign trade cargo
宁波 Ningbo	中国 China	383.9	内外贸货物 domestic and foreign trade cargo
天津 Tianjin	中国 China	381.1	内外贸货物 domestic and foreign trade cargo
广州 Guangzhou	中国 China	364.0	内外贸货物 domestic and foreign trade cargo
青岛 Qingdao	中国 China	315.5	内外贸货物 domestic and foreign trade cargo
秦皇岛 Qinhuangdao	中国 China	249.4	内外贸货物 domestic and foreign trade cargo
香港 Hong Kong	中国 China	243.0	外贸货物 foreign trade cargo
南路易斯安娜 Southern Louisiana	美国 United States	226.7	内外贸货物 domestic and foreign trade cargo
釜山 Pusan	韩国 Republic of Korea	226.2	内外贸货物 domestic and foreign trade cargo
休斯敦 Houston	美国 United States	204.6	外贸货物 foreign trade cargo
大连 Dalian	中国 China	272.0	内外贸货物 domestic and foreign trade cargo
深圳 Shenzhen	中国 China	193.7	内外贸货物 domestic and foreign trade cargo
黑德兰港 Headland Harbour	澳大利亚 Australia	178.6	内外贸货物 domestic and foreign trade cargo
光阳 Gwang	韩国 Republic of Korea	176.5	内外贸货物 domestic and foreign trade cargo
蔚山 Ulsan	韩国 Republic of Korea	170.3	内外贸货物 domestic and foreign trade cargo
名古屋 Nagoya	日本 Japan	165.1	内外贸货物 domestic and foreign trade cargo
安特卫普 Antwerp	比利时 Belgium	157.8	内外贸货物 domestic and foreign trade cargo
洛杉矶 Los Angeles	美国 United States	157.4	内外贸货物 domestic and foreign trade cargo

资料来源：Shipping Statistics and Market Review, December 2009, ISL.
Source: Shipping Statistics and Market Review, December 2009, ISL.

11-20 海上商船拥有量居世界前20位的国家或地区（2009年）
World Top Twenty Countries or Regions in Terms of the Number of Maritime Merchant Ships owned, 2009

国家和地区 Country and Region	艘数（艘）Number of Vessels (unit)	载重吨 Deadweight ton	
		万吨 10 000 tons	占世界% Percentage in the World Total
世界总计 World Total	**32 742**	**113 911.0**	**100.0**
希腊 Greece	3 120	18 753.8	16.5
日本 Japan	3 668	18 319.3	16.1
德国 Germany	3 567	10 412.9	9.1
中国 China	3 212	10 186.6	8.9
韩国 Republic of Korea	1 121	4 430.2	3.9
挪威 Norway	1 461	3 774.2	3.3
中国香港 Hong Kong, China	655	3 564.0	3.1
美国 United States	916	3 486.7	3.1
丹麦 Denmark	843	3 220.2	2.8
英国 United Kingdom	661	3 207.9	2.8
新加坡 Singapore	809	3 110.9	2.7
中国台湾省 China's Taiwan Province	624	2 993.3	2.6
意大利 Italy	755	2 103.8	1.8
俄罗斯 Russian Fed.	1 370	1 848.8	1.6
土耳其 Turkey	1 225	1 720.1	1.5
百慕大 Bermuda	151	1 710.1	1.5
印度 India	328	1 657.5	1.5
加拿大 Canada	310	1 576.2	1.4
伊朗 Iran	155	1 399.0	1.2
沙特阿拉伯 Saudi Arabia	113	1 398.5	1.2

注：1. 表中数据按载重吨排序。
2. 统计范围为1000总吨及以上船舶，截止日期均为2010年1月1日。
3. 因统计口径不同，表中数据与其他出版物公布的数据稍有差别。

Note: 1. The data in the table are arranged in the order of deadweight tons.
2. The ships, ranging over 1000 tons and more in gross ton, are counted as of Jan. 1, 2010.
3. Owing to different statistical requirements, the data given in the table may be somewhat different from those issued in other publications.

11-21 集装箱船拥有量居世界前20位的国家或地区（2009年）
World Top Twenty Countries or Regions in Terms of the number of Container Ships Owned, 2009

国家和地区 Country and Region	艘数 （艘） Number of Vessels (unit)	集装箱位 Container	
		万标准箱 10 000 TEU	占世界 % Percentage in the World
合计 Total	**4 205**	**1 222.9**	**95.4**
巴拿马 Panama	751	260.5	20.3
利比里亚 Liberia	799	256.5	20.0
德国 Germany	323	120.6	9.4
英国 United Kingdom	216	81.3	6.3
中国香港 Hong Kong, China	246	78.7	6.1
新加坡 Singapore	304	76.7	6.0
安提瓜和巴布亚 Antigua and Papua	405	56.3	4.4
丹麦 Denmark	86	47.0	3.7
马绍尔群岛 Marshall Islamds	195	40.7	3.2
塞浦路斯 Cyprus	196	38.2	3.0
中国大陆 China mainland	189	35.9	2.8
美国 United States	87	27.5	2.1
马耳他 Malta	83	21.5	1.7
希腊 Greece	35	18.7	1.5
荷兰 Netherlands	78	15.2	1.2
法国 France	25	14.5	1.1
巴哈马 Bahamas	54	13.1	1.0
意大利 Italy	22	8.2	0.6
韩国 Republic of Korea	68	6.0	0.5
马来西亚 Malaysia	43	5.8	0.5

注：1. 表中数据按载重吨排序。
2. 本表统计范围为300总吨及以上船舶，截止日期均为2010年1月1日。
3. 因统计口径不同，表中数据与其他出版物公布的数据稍有差别。

Note: 1. The data in the table are arranged in the order of deadweight tons.
2. Ships, ranging over 300 tons and more in gross ton, are counted as of Jan. 1, 2010
3. Owing to different statistical requirements, the data given in the table may be somewhat different from those issued in other publications.